T0231692

High Performance Computing

Programming and Applications

Chapman & Hall/CRC
Computational Science Series

SERIES EDITOR

Horst Simon
Deputy Director
Lawrence Berkeley National Laboratory
Berkeley, California, U.S.A.

AIMS AND SCOPE

This series aims to capture new developments and applications in the field of computational science through the publication of a broad range of textbooks, reference works, and handbooks. Books in this series will provide introductory as well as advanced material on mathematical, statistical, and computational methods and techniques, and will present researchers with the latest theories and experimentation. The scope of the series includes, but is not limited to, titles in the areas of scientific computing, parallel and distributed computing, high performance computing, grid computing, cluster computing, heterogeneous computing, quantum computing, and their applications in scientific disciplines such as astrophysics, aeronautics, biology, chemistry, climate modeling, combustion, cosmology, earthquake prediction, imaging, materials, neuroscience, oil exploration, and weather forecasting.

PUBLISHED TITLES

PETASCALE COMPUTING: ALGORITHMS AND APPLICATIONS
Edited by David A. Bader

PROCESS ALGEBRA FOR PARALLEL AND DISTRIBUTED PROCESSING
Edited by Michael Alexander and William Gardner

GRID COMPUTING: TECHNIQUES AND APPLICATIONS
Barry Wilkinson

INTRODUCTION TO CONCURRENCY IN PROGRAMMING LANGUAGES
Matthew J. Sottile, Timothy G. Mattson, and Craig E Rasmussen

INTRODUCTION TO SCHEDULING
Yves Robert and Frédéric Vivien

SCIENTIFIC DATA MANAGEMENT: CHALLENGES, TECHNOLOGY, AND DEPLOYMENT
Edited by Arie Shoshani and Doron Rotem

INTRODUCTION TO THE SIMULATION OF DYNAMICS USING SIMULINK®
Michael A. Gray

INTRODUCTION TO HIGH PERFORMANCE COMPUTING FOR SCIENTISTS
AND ENGINEERS, **Georg Hager and Gerhard Wellein**

PERFORMANCE TUNING OF SCIENTIFIC APPLICATIONS, **Edited by David Bailey,
Robert Lucas, and Samuel Williams**

HIGH PERFORMANCE COMPUTING: PROGRAMMING AND APPLICATIONS
John Levesque with Gene Wagenbreth

High Performance Computing

Programming and Applications

John Levesque

with Gene Wagenbreth

CRC Press
Taylor & Francis Group
Boca Raton London New York

CRC Press is an imprint of the
Taylor & Francis Group an **informa** business
A CHAPMAN & HALL BOOK

Chapman & Hall/CRC
Taylor & Francis Group
6000 Broken Sound Parkway NW, Suite 300
Boca Raton, FL 33487-2742

© 2011 by Taylor and Francis Group, LLC
Chapman & Hall/CRC is an imprint of Taylor & Francis Group, an Informa business

No claim to original U.S. Government works

Printed in the United States of America on acid-free paper
10 9 8 7 6 5 4 3 2 1

International Standard Book Number: 978-1-4200-7705-6 (Hardback)

Library of Congress Cataloging-in-Publication Data

Levesque, John M.
   High performance computing : programming and applications / John Levesque, Gene Wagenbreth.
      p. cm. -- (Chapman & Hall/CRC computational science series)
   Includes bibliographical references and index.
   ISBN 978-1-4200-7705-6 (hardcover : alk. paper)
   1. High performance computing. 2. Supercomputers--Programming. I. Wagenbreth, Gene. II. Title.

QA76.88.L48 2011
004.1'1--dc22   2010044861

Visit the Taylor & Francis Web site at
http://www.taylorandfrancis.com

and the CRC Press Web site at
http://www.crcpress.com

Contents

Introduction, xi

CHAPTER 1 ■ Multicore Architectures   1

  1.1   MEMORY ARCHITECTURE   1

         1.1.1   Why the Memory Wall?   2

         1.1.2   Hardware Counters   3

         1.1.3   Translation Look-Aside Buffer   4

         1.1.4   Caches   6

                  1.1.4.1   Associativity   6

                  1.1.4.2   Memory Alignment   11

         1.1.5   Memory Prefetching   11

  1.2   SSE INSTRUCTIONS   12

  1.3   HARDWARE DESCRIBED IN THIS BOOK   14

         EXERCISES   16

CHAPTER 2 ■ The MPP: A Combination of Hardware
            and Software   19

  2.1   TOPOLOGY OF THE INTERCONNECT   20

         2.1.1   Job Placement on the Topology   21

  2.2   INTERCONNECT CHARACTERISTICS   22

         2.2.1   The Time to Transfer a Message   22

         2.2.2   Perturbations Caused by Software   23

  2.3   NETWORK INTERFACE COMPUTER   24

  2.4   MEMORY MANAGEMENT FOR MESSAGES   24

2.5 HOW MULTICORES IMPACT THE
PERFORMANCE OF THE INTERCONNECT   25

EXERCISES   25

CHAPTER 3 ■ How Compilers Optimize Programs   27

3.1 MEMORY ALLOCATION   27

3.2 MEMORY ALIGNMENT   28

3.3 VECTORIZATION   29

   3.3.1 Dependency Analysis   31

   3.3.2 Vectorization of IF Statements   32

   3.3.3 Vectorization of Indirect Addressing and Strides   34

   3.3.4 Nested DO Loops   35

3.4 PREFETCHING OPERANDS   36

3.5 LOOP UNROLLING   37

3.6 INTERPROCEDURAL ANALYSIS   38

3.7 COMPILER SWITCHES   38

3.8 FORTRAN 2003 AND ITS INEFFICIENCIES   39

   3.8.1 Array Syntax   40

   3.8.2 Using Optimized Libraries   42

   3.8.3 Passing Array Sections   42

   3.8.4 Using Modules for Local Variables   43

   3.8.5 Derived Types   43

3.9 SCALAR OPTIMIZATIONS PERFORMED
BY THE COMPILER   44

   3.9.1 Strength Reduction   44

   3.9.2 Avoiding Floating Point Exponents   46

   3.9.3 Common Subexpression Elimination   47

EXERCISES   48

CHAPTER 4 ■ Parallel Programming Paradigms   51

4.1 HOW CORES COMMUNICATE WITH EACH OTHER   51

   4.1.1 Using MPI across All the Cores   51

4.1.2   Decomposition   52

4.1.3   Scaling an Application   53

4.2   MESSAGE PASSING INTERFACE   55

4.2.1   Message Passing Statistics   55

4.2.2   Collectives   56

4.2.3   Point-to-Point Communication   57

4.2.3.1   Combining Messages into
          Larger Messages   58

4.2.3.2   Preposting Receives   58

4.2.4   Environment Variables   61

4.2.5   Using Runtime Statistics to Aid MPI-Task
        Placement   61

4.3   USING OPENMP™   62

4.3.1   Overhead of Using OpenMP™   63

4.3.2   Variable Scoping   64

4.3.3   Work Sharing   66

4.3.4   False Sharing in OpenMP™   68

4.3.5   Some Advantages of Hybrid Programming:
        MPI with OpenMP™   70

4.3.5.1   Scaling of Collectives   70

4.3.5.2   Scaling Memory Bandwidth Limited MPI
          Applications   70

4.4   POSIX® THREADS   71

4.5   PARTITIONED GLOBAL ADDRESS SPACE
      LANGUAGES (PGAS)   77

4.5.1   PGAS for Adaptive Mesh Refinement   78

4.5.2   PGAS for Overlapping Computation and
        Communication   78

4.5.3   Using CAF to Perform Collective Operations   79

4.6   COMPILERS FOR PGAS LANGUAGES   83

4.7   ROLE OF THE INTERCONNECT   85

EXERCISES   85

CHAPTER 5 ■ A Strategy for Porting an Application
to a Large MPP System   87

5.1   GATHERING STATISTICS FOR A LARGE
       PARALLEL PROGRAM   89

EXERCISES   96

CHAPTER 6 ■ Single Core Optimization   99

6.1   MEMORY ACCESSING   99

       6.1.1   Computational Intensity   102

       6.1.2   Looking at Poor Memory Utilization   103

6.2   VECTORIZATION   109

       6.2.1   Memory Accesses in Vectorized Code   110

       6.2.2   Data Dependence   112

       6.2.3   IF Statements   119

       6.2.4   Subroutine and Function Calls   129

               6.2.4.1   Calling Libraries   135

       6.2.5   Multinested DO Loops   136

               6.2.5.1   More Efficient Memory Utilization   145

6.3   SUMMARY   156

       6.3.1   Loop Reordering   156

       6.3.2   Index Reordering   157

       6.3.3   Loop Unrolling   157

       6.3.4   Loop Splitting (Loop Fission)   157

       6.3.5   Scalar Promotion   157

       6.3.6   Removal of Loop-Independent IFs   158

       6.3.7   Use of Intrinsics to Remove IFs   158

       6.3.8   Strip Mining   158

       6.3.9   Subroutine Inlining   158

       6.3.10 Pulling Loops into Subroutines   158

       6.3.11 Cache Blocking   159

       6.3.12 Loop Jamming (Loop Fusion)   159

EXERCISES   159

Chapter 7 ■ Parallelism across the Nodes   161

7.1   LESLIE3D   162

7.2   PARALLEL OCEAN PROGRAM (POP)   164

7.3   SWIM   169

7.4   S3D   171

7.5   LOAD IMBALANCE   174

    7.5.1   SWEEP3D   177

7.6   COMMUNICATION BOTTLENECKS   179

    7.6.1   Collectives   179

        7.6.1.1   Writing One's Own Collectives to
                Achieve Overlap with Computation   179

    7.6.2   Point to Point   180

    7.6.3   Using the "Hybrid Approach" Efficiently   180

        7.6.3.1   SWIM Benchmark   181

7.7   OPTIMIZATION OF INPUT AND OUTPUT (I/O)   182

    7.7.1   Parallel File Systems   183

    7.7.2   An Inefficient Way of Doing I/O on a Large
           Parallel System   184

    7.7.3   An Efficient Way of Doing I/O on a Large
           Parallel System   184

EXERCISES   185

Chapter 8 ■ Node Performance   187

8.1   WUPPERTAL WILSON FERMION SOLVER:
      WUPWISE   187

8.2   SWIM   189

8.3   MGRID   192

8.4   APPLU   192

8.5   GALGEL   194

8.6   APSI   196

8.7   EQUAKE   199

8.8   FMA-3D   201

8.9   ART   203

8.10 ANOTHER MOLECULAR MECHANICS
PROGRAM (AMMP)   204

8.11 SUMMARY   206

EXERCISES   207

CHAPTER 9 ■ Accelerators and Conclusion   209

9.1   ACCELERATORS   210

9.1.1 Using Extensions to OpenMP™ Directives
       for Accelerators   211

9.1.2 Efficiency Concerns When Using an Accelerator 212

9.1.3 Summary for Using Accelerators   213

9.1.4 Programming Approaches for Using Multicore
       Nodes with Accelerators   214

9.2   CONCLUSION   214

EXERCISES   215

APPENDIX A: COMMON COMPILER DIRECTIVES, 217

APPENDIX B: SAMPLE MPI ENVIRONMENT VARIABLES, 219

REFERENCES, 221

INDEX, 223

Introduction

The future of high-performance computing (HPC) lies with large distributed parallel systems with three levels of parallelism, thousands of nodes containing MIMD* groups of SIMD* processors. For the past 50 years, the clock cycle of single processors has decreased steadily and the cost of each processor has decreased. Most applications would obtain speed increases inversely proportional to the decrease in the clock cycle and cost reduction by running on new hardware every few years with little or no additional programming effort. However, that era is over. Users now must utilize parallel algorithms in their application codes to reap the benefit of new hardware advances.

In the near term, the HPC community will see an evolution in the architecture of its basic building block processors. So, while clock cycles are not decreasing in these newer chips, we see that Moore's law still holds with denser circuitry and multiple processors on the die. The net result is more floating point operations/second (FLOPS) being produced by wider functional units on each core and multiple cores on the chip. Additionally, a new source of energy-efficient performance has entered the HPC community. The traditional graphics processing units are becoming more reliable and more powerful for high-precision operations, making them viable to be used for HPC applications.

But to realize this increased potential FLOP rate from the wider instruction words and the parallel vector units on the accelerators, the compiler must generate vector or streaming SIMD extension (SSE) instructions. Using vector and SSE instructions is by no means automatic. The user must either hand code the instructions using the assembler and/or the compiler must generate the SSE instructions and vector code when compiling the program. This means that the compiler must perform the

---

* MIMD—multiple instruction, multiple data; SIMD—single instruction, multiple data.

extra dependence analysis needed to transform scalar loops into vectorizable loops. Once the SSE instructions on the X86 processor are employed, a significant imbalance between greater processor performance and limited memory bandwidth is revealed. Memory architectures may eventually remove this imbalance, but in the near term, poor memory performance remains as a major concern in the overall performance of HPC applications.

Moving beyond the chip, clustered systems of fast multicore processors and/or accelerators on the node immediately reveal a serious imbalance between the speed of the network and the performance of each node. While node processor performance gains are from multiple cores, multiple threads, and multiple wide functional units, the speed gains of network interconnects lag and are not as likely to increase at the same rate to keep up with the nodes. Programmers will have to rethink how they use those high-performing nodes in a parallel job, because this recent chip development with multiple cores per node demands a more efficient message passing code.

This book attempts to give the HPC applications programmer an awareness of the techniques needed to address these new performance issues. We discuss hardware architectures from the point of view of what an application developer really needs to know to achieve high performance, leaving out the deep details of how the hardware works. Similarly, when discussing programming techniques to achieve performance, we avoid detailed discourse on programming language syntax or semantics other than just what is required to get the idea behind those techniques across. In this book, we concentrate on C and Fortran, but the techniques described may just as well be applicable to other languages such as C++ and JAVA®.

Throughout the book, the emphasis will be on chips from Advanced Micro Devices (AMD) and systems, and interconnects and software from Cray Inc., as the authors have much more experience on those systems. While the concentration is on a subset of the HPC industry, the techniques discussed have application across the entire breath of the HPC industry, from the desktop to the Petaflop system. Issues that give rise to bottlenecks to attaining good performance will be discussed in a generic sense, having applicability to all vendors' systems.

To set the foundation, users must start thinking about three levels of parallelism. At the outermost level is message passing to communicate between the nodes of the massively parallel computer, shared memory parallelism in the middle to utilize the cores on the nodes or the MIMD units on the accelerator, and finally, vectorization on the inner level. Techniques to address each of these levels are found within these pages.

What are the application programmers to do when asked to port their application to a massively parallel multicore architecture? After discussing the architectural and software issues, the book outlines a strategy for taking an existing application and identifying how to address the formidable task of porting and optimizing their application for the target system. The book is organized to facilitate the implementation of that strategy. First, performance data must be collected for the application and the user is directed to individual chapters for addressing the bottlenecks indicated in the performance data. If the desire is to scale the application to 10,000 processors and it currently quits scaling at 500 processors, processor performance is not important, but improving the message passing and/or reducing synchronization time or load imbalance is what should be addressed. If, on the other hand, the application scales extremely well, only 1–2% of peak performance is being seen, processor performance should be examined.

Chapter 1 addresses the node architecture in the latest chips from AMD. AMD supplies the same basic chip instruction architecture as Intel®, which can be employed as single-socket nodes, or as cache-coherent, multisocket nodes. Of particular interest here is the memory architecture, with its caches and prefetching mechanisms, and the high-performance functional units capable of producing four 8-byte or eight 4-byte floating point results/core. And, as mentioned earlier, some sort of "vectorization" scheme must be employed to achieve these high result rates. While the unit of computation will primarily be the core, the cores are organized within a uniform memory shared memory socket, and then there are multiple sockets per node. Within the socket, the cores share memory with uniform memory access (UMA); that is, fetching any memory on the socket from any core on the socket will take the same amount of time. Across the sockets, the cores experience a nonuniform memory access (NUMA); that is, the time to access memory connected to a particular core's socket is faster than accessing memory connected to the other socket. The latency is higher and the bandwidth is slightly smaller.

Chapter 2 discusses the infrastructure of connecting the nodes together including hardware and software, the hardware that supplies connectivity between the nodes in a massively parallel processor (MPP) system and the software that orchestrates the MPP to work on a single application. Certain characteristics of the interconnect such as latency, injection bandwidth, global bandwidth, and messages/second that can be handled by the network interface computer (NIC) play an important role in how well the application scales across the nodes in the system.

Chapter 3 brings the role of compilers into discussion. In this chapter, we concentrate on the compilers from Portland Group Inc. (PGI) and the Cray compilation environment (CCE). Users tend to expect (or hope) that the compiler will automatically perform all the right transformations and optimizations and generate the best possible executable code. Unfortunately, reality is far from this; however, it is not entirely the compiler technology at fault, but the way the application is written and the ambiguities inherent in the source code. Typically, what the compiler needs to know to properly transform and optimize the generated executable code is hidden, and so it must take a conservative approach by making minimal assumptions in order to preserve the validity of the program. This chapter discusses various source code syndromes that inhibit compilers from optimizing the code. We will see that, for some compilers, simple comment line directives can help remove certain optimization inhibitors.

In Chapter 4, we discuss the message passing interface (MPI) and OpenMP™, which are the most prevalent parallel programming approaches used today and continue with Pthreads for shared memory parallelization and the partitioned global address space (PGAS) languages co-array Fortran and unified parallel C (UPC). We focus primarily on MPI message passing and OpenMP shared memory models, and the hybrid model of distributed shared memory (DSM) that applies both MPI and OpenMP together over thousands of multicore nodes. For the hybrid DSM approach to be successful, both the MPI calls across the network, and the placement of OpenMP directives on the shared memory node must be done effectively. We show in Chapter 4 the high-level mechanics of how this might be achieved. In later chapters, specific applications that have been implemented in OpenMP and MPI are discussed to see the good and not-so-good implementations.

Given the discussion in the first four chapters, we then propose a strategy to be taken by an application developer when faced with the challenge of porting and optimizing their application for the multicore, massively parallel architecture. This strategy starts with the process of gathering runtime statistics from running the application on a variety of processor configurations. Chapter 5 discusses what statistics would be required and how one should interpret the data and then proceed to investigate the optimization of the application. Given the runtime statistics, the user should address the most obvious bottleneck first and work on the "whack a mole principal"; that is, address the bottlenecks from the highest to the lowest. Potentially, the highest bottleneck could be single-processor performance, single-node performance, scaling issues, or input/output (I/O).

If the most time-consuming bottleneck is single-processor performance, Chapter 6 discusses how the user can address performance bottlenecks in processor or single-core performance. In this chapter, we examine computational kernels from a wide spectrum of real HPC applications. Here readers should find examples that relate to important code sections in their own applications. The actual timing results for these examples were gathered by running on hardware current at the time this book was published. The companion Web site (www.hybridmulticoreoptimization.com) contains all the examples from the book, along with updated timing results on the latest released processors. The focus here is on single-core performance, efficient cache utilization, and loop vectorization techniques.

Chapter 7 addresses scaling performance. If the most obvious bottleneck is communication and/or synchronization time, the user must understand how a given application can be restructured to scale to higher processor counts. This chapter looks at numerous MPI implementations, illustrating bottlenecks to scaling. By looking at existing message passing implementations, the reader may be able to identify those techniques that work very well and some that have limitations in scaling to a large number of processors.

In Chapter 7, we address the optimization of I/O as an application is moved to higher processor counts. Of course, this section depends heavily on the network and I/O capabilities of the target system; however, there is a set of rules that should be considered on how to do I/O from a large parallel job.

As the implementation of a hybrid MPI/OpenMP programming paradigm is one method of improving scalibility of an application, Chapter 8 is all about OpenMP performance issues. In particular, we discuss the efficient use of OpenMP by looking at the Standard Performance Evaluation Corporation (SPEC®) OpenMP examples. In this chapter, we look at each of the applications and discuss memory bandwidth issues that arise from multiple cores accessing memory simultaneously. We see that efficient OpenMP requires good cache utilization on each core of the node.

A new computational resource is starting to become a viable HPC alternative. General purpose graphics processor units (GPGPUs) have gained significant market share in the gaming and graphics area. These systems previously did not have an impact on the HPC community owing to the lack of hardware 64-bit operations. The performance of the high-precision software was significantly lower than the hardware 32-bit performance. Initially, the GPGPUs did not have error correction, because it was really not required by the gaming industry. For these reasons, the original releases of these systems were not viable for HPC where high precision and reliability are required. Recently, this situation has changed. Once again, with the ability to pack more and more into the chip, GPGPUs are being designed with hardware 64-bit precision whose performance is close to the 32-bit performance (2–3 instead of 10–12 difference) and error detection–correction is being introduced to enable correction of single-bit errors and detection of double-bit errors. These accelerators supply impressive FLOPS/clock by supplying multiple SIMD units. The programming paradigm is therefore to parallelize the outermost loops of a kernel as in OpenMP or threading and vectorize the innermost loops. Unlike the SSE instructions, vectorization for the accelerators can deliver factors of 10–20 in performance improvement. The final chapter will look at the future and discuss what the application programmer should know about utilizing the GPGPUs to carry out HPC.

CHAPTER 1

Multicore Architectures

Tᴴᴇ ᴍᴜʟᴛɪᴄᴏʀᴇ ᴀʀᴄʜɪᴛᴇᴄᴛᴜʀᴇꜱ that we see today are due to an increased desire to supply more floating point operations (FLOPS) each clock cycle from the computer chip. The typical decrease of the clock cycle that we have seen over the past 20 years is no longer evident; in fact, on some systems the clock cycles are becoming longer and the only way to supply more FLOPS is by increasing the number of cores on the chip or by increasing the number of floating point results from the functional units or a combination of the two.

When the clock cycle shortens, everything on the chip runs faster without programmer interaction. When the number of results per clock cycle increases, it usually means that the application must be structured to optimally use the increased operation count. When more cores are introduced, the application must be restructured to incorporate more parallelism. In this section, we will examine the most important elements of the multicore system—the memory architecture and the vector instructions which supply more FLOPS/clock cycle.

1.1 MEMORY ARCHITECTURE

The faster the memory circuitry, the more the memory system costs. Memory hierarchies are built by having faster, more-expensive memory close to the CPU and slower, less-expensive memory further away. All multicore architectures have a memory hierarchy: from the register set, to the various levels of cache, to the main memory. The closer the memory is to the processing unit, the lower the latency to access the data from that memory component and the higher the bandwidth for accessing the data.

For example, registers can typically deliver multiple operands to the functional units in one clock cycle and multiple registers can be accessed in each clock cycle. Level 1 cache can deliver 1–2 operands to the registers in each clock cycle. Lower levels of cache deliver fewer operands per clock cycle and the latency to deliver the operands increases as the cache is further from the processor. As the distance from the processor increases the size of the memory component increases. There are 10s of registers, Level 1 cache is typically 64 KB (when used in this context, K represents 1024 bytes—64 KB is 65,536 bytes). Higher levels of cache hold more data and main memory is the largest component of memory.

Utilizing this memory architecture is the most important lesson a programmer can learn to effectively program the system. Unfortunately, compilers cannot solve the memory locality problem automatically. The programmer must understand the various components of the memory architecture and be able to build their data structures to most effectively mitigate the lack of sufficient memory bandwidth.

The amount of data that is processed within a major computational routine must be known to understand how that data flows from memory through the caches and back to memory. To some extent the computation performed on the data is not important. A very important optimization technique, "cache blocking," is a restructuring process that restricts the amount of data processed in a computational chunk so that the data fits within Level 1 and/or Level 2 cache during the execution of the blocked DO loop. Cache blocking will be discussed in more detail in Chapter 6.

1.1.1 Why the Memory Wall?

On today's, as well as future generation, microprocessors the ratio between the rate at which results can be generated by the functional units, result rate, divided by the rate at which operands can be fetched from and stored to memory is becoming very large. For example, with the advent of the SSE3 (streaming single instruction, multiple data extensions) instruction the result rate doubled and the memory access rate remained the same. Many refer to the "memory wall" as a way of explaining that memory technologies have not kept up with processor technology and are far behind the processor in producing inexpensive fast circuitry.

> We all know that the rate of improvement in the microprocessor speed exceeds the rate of improvement in DRAM memory speed, each is improving exponentially, but the exponent for micro-processors is

substantially larger than that for DRAMs. The difference between diverging exponentials also grows exponentially; so, although the disparity between processor and memory speed is already an issue, downstream someplace it will be a much bigger one.

Wulf and McKee, 1995

Given this mismatch between processor speed and memory speed, hardware designers have built special logic to mitigate this imbalance. This section explains how the memory system of a typical microprocessor works and what the programmer must understand to allocate their data in a form that can most effectively be accessed during the execution of their program. Understanding how to most effectively use memory is perhaps the most important lesson a programmer can learn to write efficient programs.

1.1.2 Hardware Counters

Since the early days of super scalar processing, chip designers incorporated hardware counters into their systems. These counters were originally introduced to measure important performance issues in the hardware. HPC proponents soon found out about these counters and developed software to read the counters from an application program. Today, software supplied by the Innovative Computing Laboratory at the University of Tennessee called PAPI, gives "raw" hardware counter data [2]. Throughout the remainder of the book, the importance of this information will be discussed. Within the book, the CrayPat™ performance analysis tool [3] will be used to access the hardware counters for examples shown. Most vendors supply profiling software that can access these hardware counters.

It is important to understand that the "raw" hardware counter data are somewhat useless. For example, when one runs an application the number of cache misses is not enough information. What is needed is the number of memory accesses for each cache miss. This important piece of information is known as a derived metric. It is obtained from the performance tools measuring the number of cache misses and the number of memory accesses and computing the derived metric (memory accesses/cache miss).

Without the hardware counter data the application programmer would have a very difficult time figuring out why their application did not run well. With these data, the application programmer can quickly zero into the portion of the application that takes most of the CPU time and then understand why the performance is or is not good.

In the following discussion, it is important to understand that derived metrics are available to be used in a profiling mode to determine the important memory access details about the executing application.

1.1.3 Translation Look-Aside Buffer

The mechanics of the translation look-aside buffer (TLB) is the first important lesson in how to effectively utilize memory. When the processor issues an address for a particular operand, that address is a logical address. Logical addresses are what the application references and in the applications view, consecutive logical addresses are contiguous in memory. In practice, this is not the case. Logical memory is mapped onto physical memory with the use of the TLB. The TLB contains entries that give the translation of the location of logical memory pages within physical memory (Figure 1.1). The default page size on most Linux® systems is 4096 bytes. If one stores four-byte operands, there will be 1024 four-byte operands in a page. The page holds 512 eight-byte operands.

Physical memory can be fragmented and two pages adjacent in logical memory may not be adjacent in physical memory (Figure 1.2). The mapping between the two, logical and physical, is performed by an entry in the TLB. The first time an operand is fetched to the processor, the page table

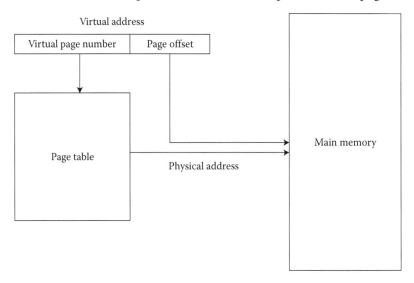

FIGURE 1.1 Operation of the TLB to locate operand in memory. (Adapted from Hennessy, J. L. and Patterson, D. A. with contributions by Dusseau, A. C. et al. *Computer Architecture—A Quantitative Approach*. Burlington, MA: Morgan Kaufmann publications.)

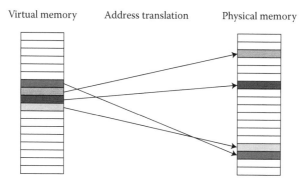

Virtual memory   Address translation   Physical memory

FIGURE 1.2   Fragmentation in physical memory.

entry must be fetched, the logical address translated to a physical address and then the cache line that contains the operand is fetched to Level 1 cache. To access a single element of memory, two memory loads, the table entry and the cache line, are issued and 64 bytes are transferred to Level 1 cache—all this for supplying a single operand. If the next operand is contiguous to the previous one in logical memory, the likelihood of it being in the same page and same cache line is very high. (The only exception would be if the first operand was the last operand in the cache line which happens to be the last cache line in the page).

A TLB table entry allows the processor to access all the 4096 bytes within the page. As the processor accesses additional operands, either the operand resides within a page whose address resides in the TLB or another table entry must be fetched to access the physical page containing the operand. A very important hardware statistic that can be measured for a section of code is the effective TLB miss ratio. A TLB miss is the term used to describe the action when no table entry within the TLB contains the physical address required—then a miss occurs and a page entry must be loaded to the TLB and then the cache line that contains the operands can be fetched. A TLB miss is very expensive, since it requires two memory accesses to obtain the data. Unfortunately, the size of the TLB is relatively small. On some AMD microprocessors the TLB only has 48 entries, which means that the maximum amount of memory that can be "mapped" at any given time is $4096 \times 48 = 196,608$ bytes. Given this limitation, the potential of TLB "thrashing" is possible. Thrashing the TLB refers to the condition where very few of the operands within the page are referenced before the page table entry is flushed from the TLB. In subsequent chapters, examples where the TLB performance is poor will be discussed in more detail. The programmer should always obtain hardware counters for the important sections of their application to determine if their TLB utilization is a problem.

1.1.4 Caches

The processor cannot fetch a single bit, byte, word it must fetch an entire cache line. Cache lines are 64 contiguous bytes, which would contain either 16 4-byte operands or 8 8-byte operands. When a cache line is fetched from memory several memory banks are utilized. When accessing a memory bank, the bank must refresh. Once this happens, new data cannot be accessed from that bank until the refresh is completed. By having interleaved memory banks, data from several banks can be accessed in parallel, thus delivering data to the processor at a faster rate than any one memory bank can deliver. Since a cache line is contiguous in memory, it spans several banks and by the time a second cache line is accessed the banks have had a chance to recycle.

1.1.4.1 Associativity

A cache is a highly structured, expensive collection of memory banks. There are many different types of caches. A very important characteristic of the cache is cache associativity. In order to discuss the operation of a cache we need to envision the following memory structure. Envision that memory is a two dimensional grid where the width of each box in the grid is a cache line. The size of a row of memory is the same as one associativity class of the cache. Each associativity class also has boxes the size of a cache line. When a system has a direct mapped cache, this means that there is only one associativity class. On all x86 systems the associativity of Level 1 cache is two way. This means that there are two rows of associativity classes in Level 1 cache. Now consider a column of memory. Any cache line in a given column of memory must be fetched to the corresponding column of the cache. In a two-way associative cache, there are only two locations in Level 1 cache for any of the cache lines in a given column of memory. The following diagram tries to depict the concept.

Figure 1.3 depicts a two-way associative Level 1 cache. There are only two cache lines in Level 1 cache that can contain any of the cache lines in the Nth column of memory. If a third cache line from the same column is required from memory, one of the cache lines contained in associativity

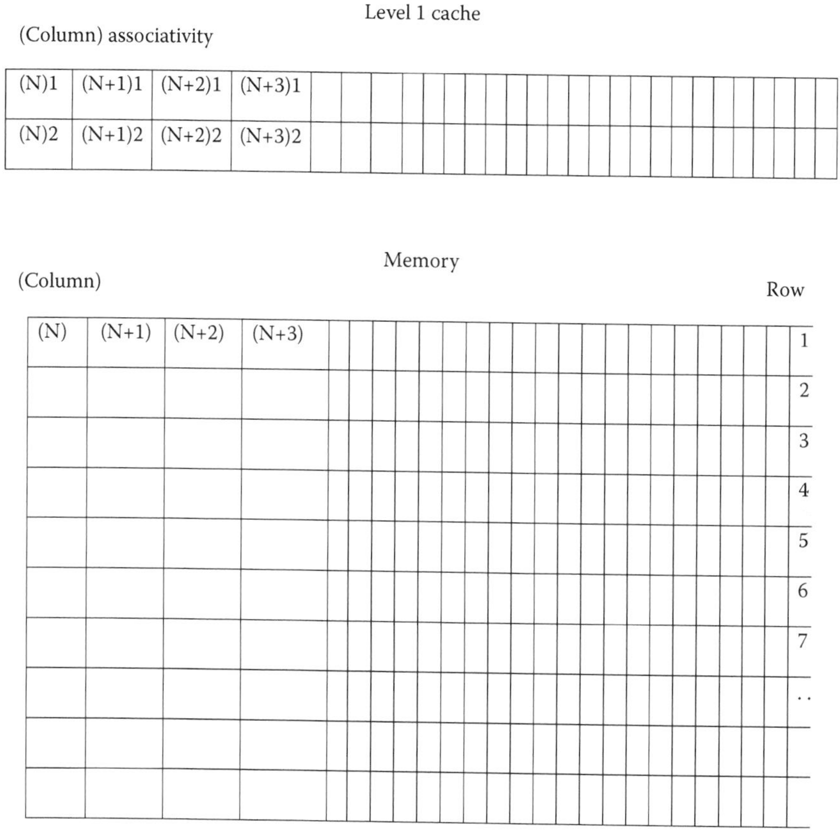

(Column) associativity

Level 1 cache

(N)1	(N+1)1	(N+2)1	(N+3)1																					
(N)2	(N+1)2	(N+2)2	(N+3)2																					

(Column)

Memory

Row

(N)	(N+1)	(N+2)	(N+3)																		
																					1
																					2
																					3
																					4
																					5
																					6
																					7
																					..

FIGURE 1.3   Two-way associative level 1 cache.

class 1 or associativity class 2 must be flushed out of cache, typically to Level 2 cache. Consider the following DO loop:

```
REAL*8 A(65536),B(65536),C(65536)
DO I=1,65536
      C(I) = A(I)+SCALAR*B(I)
ENDDO
```

Let us assume that A(1)–A(8) are contained in the Nth column in memory. What is contained in the second row of the Nth column? The width of the cache is the size of the cache divided by its associativity. The size of this Level 1 cache is 65,536 bytes, and each associativity class has 32,768 bytes of locations. Since the array A contains 8 bytes per word, the length of the

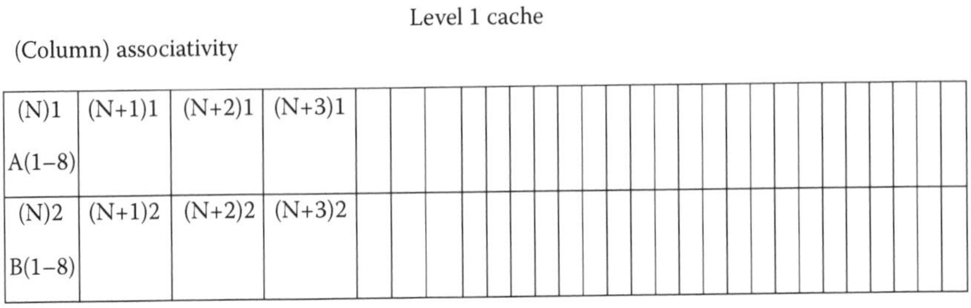

Level 1 cache

(Column) associativity

(N)1	(N+1)1	(N+2)1	(N+3)1															
A(1–8)																		
(N)2	(N+1)2	(N+2)2	(N+3)2															
B(1–8)																		

FIGURE 1.4   Contents of level 1 cache after fetch of A(1) and B(1).

A array is $65{,}536 \times 8 = 131{,}072$ bytes. The second row of the Nth column will contain A(4097–8192), the third A(8193–12,288), and the sixteen row will contain the last part of the array A. The 17th row of the Nth column will contain B(1)–B(4096), since the compiler should store the array B right after the array A and C(1)–C(4096) will be contained in 33rd row of the Nth column. Given the size of the dimensions of the arrays A, B, and C, the first cache line required from each array will be exactly in the same column. We are not concerned with the operand SCALAR—this will be fetched to a register for the duration of the execution of the DO loop.

When the compiler generates the fetch for A(1) the cache line containing A(1) will be fetched to either associativity 1 or associativity 2 in the Nth column of Level 1 cache. Then the compiler fetches B(1) and the cache line containing that element will go into the other slot in the Nth column of Level 1 cache, either into associativity 1 or associativity 2. Figure 1.4 illustrates the contents of Level 1 cache after the fetch of A(1) and B(1).

The add is then generated. To store the result into C(1), the cache line containing C(1) must be fetched to Level 1 cache. Since there are only two slots available for this cache line, either the cache line containing B(1) or the cache line containing A(1) will be flushed out of Level 1 cache into

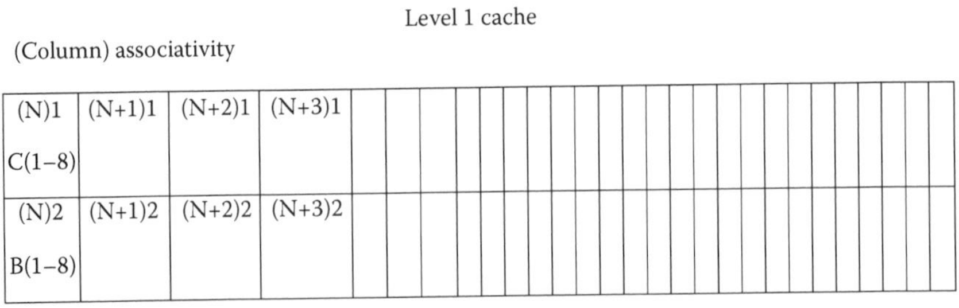

Level 1 cache

(Column) associativity

(N)1	(N+1)1	(N+2)1	(N+3)1															
C(1–8)																		
(N)2	(N+1)2	(N+2)2	(N+3)2															
B(1–8)																		

FIGURE 1.5   State of level 1 cache after fetch of C(1).

Level 2 cache. Figure 1.5 depicts the state of Level 1 cache once C(1) is fetched to cache.

The second pass through the DO loop the cache line containing A(2) or B(2) will have to be fetched from Level 2; however, the access will over-write one of the other two slots and we end up thrashing cache with the execution of this DO loop.

Consider the following storage scheme:

```
REAL*8 A(65544),B(65544),C(65544)
DO I=1,65536
      C(I) = A(I)+SCALAR*B(I)
ENDDO
```

The A array now occupies 16 full rows of memory plus one cache line. This causes B(1)–B(8) to be stored in the N + 1 column of the 17th row and C(1)–C(8) is stored in the N + 2 column of the 33rd row. We have offset the arrays and so they do not conflict in the cache. This storage therefore results in more efficient execution than that of the previous version. The next example investigates the performance impact of this rewrite. Figure 1.6 depicts the status of Level 1 cache given this new storage scheme.

Figure 1.7 gives the performance of this simple kernel for various values of vector length. In these tests the loop is executed 1000 times. This is done to measure the performance when the arrays, A, B, and C are fully con-tained within the caches. If the vector length is greater than 2730 [(65,536 bytes size of Level 1 cache)/(8 bytes for REAL(8) divided *3 operands to be fetch-stored) = 2730], the three arrays cannot be contained within Level 1 cache and we have some overflow into Level 2 cache. When the vector length is greater than 21,857 [(524,588 bytes in Level 2 cache)/(8 bytes divided by 3 operands) = 21,857] then the three arrays will not fit in Level

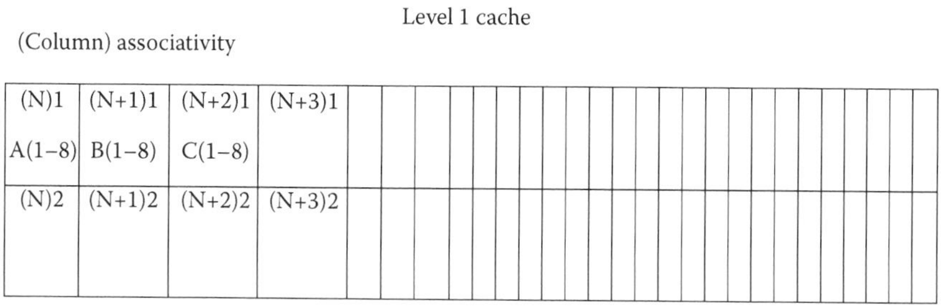

Level 1 cache

(Column) associativity

(N)1	(N+1)1	(N+2)1	(N+3)1																							
A(1–8)	B(1–8)	C(1–8)																								
(N)2	(N+1)2	(N+2)2	(N+3)2																							

FIGURE 1.6   Status of level 1 cache given offset-array storage scheme.

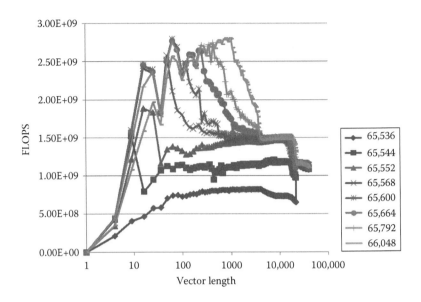

FIGURE 1.7   Performance based on the storage alignment of the arrays.

2 cache and the arrays will spill over into Level 3 cache. In Figure 1.7, there is a degradation of performance as N increases, and this is due to where the operands reside prior to being fetched.

More variation exists than that attributed to increasing the vector length. There is a significant variation of performance of this simple kernel due to the size of each of the arrays. Each individual series indicates the dimension of the three arrays in memory. Notice that the series with the worst performance is dimensioned by 65,536 REAL*8 words. The performance of this memory alignment is extremely bad because the three arrays are overwriting each other in Level 1 cache as explained earlier. The next series which represents the case where we add a cache line to the dimension of the arrays gives a slightly better performance; however, it is still poor, the third series once again gives a slightly better performance, and so on. The best-performing series is when the arrays are dimensioned by a 65,536 plus a page (512 words—8-byte words).

The reason for this significant difference in performance is due to the memory banks in Level 1 cache. When the arrays are dimensioned a large power of two they are aligned in cache, as the three arrays are accessed, the accesses pass over the same memory banks and the fetching of operands stalls until the banks refresh. When the arrays are offset by a full page of 4096 bytes, the banks have time to recycle before another operand is accessed.

The lesson from this example is to pay attention to the alignment of arrays. As will be discussed in the compiler section, the compiler can only do so much to "pad" arrays to avoid these alignment issues. The application programmer needs to understand how best to organize their arrays to achieve the best possible cache utilization. Memory alignment plays a very important role in effective cache utilization which is an important lesson a programmer must master when writing efficient applications.

1.1.4.2 Memory Alignment

A page always starts on a 4096-byte boundary. Within the page, the next unit of alignment is the cache line and there is always an even number of full cache lines within a page. A 4096-byte page would contain 64 512-bit cache lines. Within a cache line, there are 4 super words of 128 bits each. The functional units on the chips, discussed in the next section, can accept two 128 bit super words each clock cycle. Adding two arrays as in the following DO loop

```
DO I=1,20
      A(I)=B(I)+C(I)
ENDDO
```

If B(1) and C(1) are both the first elements of a super word, the move from cache to register and from register through the adder back into the register can be performed at two results a clock cycle. If on the other hand they are not aligned, the compiler must move the unaligned operands into a 128-bit register prior to issuing the add. These align operations tend to decrease the performance attained by the functional unit.

Unfortunately it is extremely difficult to assure that the arrays are aligned on super word boundaries. One way to help would be to always make the arrays a multiple of 128 bits. What about scalars? Well one can always try to pack your scalars together in the data allocation and make sure that the sum of the scalar storage is a multiple of 128 bits. If everything allocated is a length of 128 bits, the likelihood of the first word of every array being on a 128-bit boundary is assured.

1.1.5 Memory Prefetching

Memory prefetching is an optimization that both the hardware and the compiler can use to assure that operands required in subsequent iterations of the DO loop are available when the functional units are available. When the memory controller (hardware) senses a logical pattern to addressing the operands, the next logical cache line is fetched automatically before the processor requests it. Hardware prefetching cannot be turned off. The hardware does this to help mitigate the latency to access the memory.

When the compiler optimizes a DO loop, it may use software prefetches prior to the DO loop and just before the end of the DO loop to prefetch operands required in the next pass through the loop. Additionally, compilers allow the user to influence the prefetching with comment line directives. These directives give the user the ability to override the compiler and indicate which array should be prefetched.

Since two contiguous logical pages may not be contiguous in physical memory, neither type of prefetching can prefetch across a page boundary. When using small pages prefetching is limited to 63 cache lines on each array.

When an application developer introduces their own cache blocking, it is usually designed without considering hardware or software prefetching. This turns out to be extremely important since the prefetching can utilize cache space. When cache blocking is performed manually it might be best to turn off software prefetching.

The intent of this section is to give the programmer the important aspects of memory that impact the performance of their application. Future sections will examine kernels of real applications that utilize memory poorly; unfortunately most applications do not effectively utilize memory. By using the hardware counter information and following the techniques discussed in later chapters, significant performance gains can be achieved by effectively utilizing the TLB and cache.

1.2 SSE INSTRUCTIONS

On AMD's multicore chips, there are two separate functional units, one for performing an ADD and one for performing a MULTIPLY. Each of these is independent and the size of the floating point add or multiply determines how many operations can be produced in parallel. On the dual core systems, the width of the SSE instruction is 64 bits wide. If the application uses 32 bit or single precision arithmetic, each functional unit (ADD and MULTIPLY) produces two single precision results each clock cycle. With the advent of the SSE3 instructions on the quad core systems, the functional units are 128 bits wide and each functional unit produces four single precision results each clock cycle (cc) or two double precision results.

When running in REAL*8 or double precision, the result rate of the functional units is cut in half, since each result requires 64 bit wide operands. The following table gives the results rate per clock cycle (cc) from the SSE2 and SSE3 instructions.

	SSE2 (Dual Core Systems)	SSE3(Quad Core Systems)
32-bit single precision	2 Adds and 2 Multiplies/cc	4 Adds and 4 Multiplies/cc
64-bit double precision	1 Add and 1 Multiply/cc	2 Adds and 2 Multiplies/cc

For the compiler to utilize SSE instructions, it must know that the two operations being performed within the instruction are independent of each other. All the current generation compilers are able to determine the lack of dependency by analyzing the DO loop for vectorization. Vectorization will be discussed in more detail in later chapters; however, it is important to understand that the compiler must be able to vectorize a DO loop or at least a part of the DO loop in order to utilize the SSE instructions.

Not all adds and multiplies would use the SSE instructions. There is an added requirement for using the SSE instructions. The operands to the instruction must be aligned on instruction width boundary (super words of 128 bits). For example:

```
DO I=1, 100
      A(I)=B(I)+SCALAR*C(I)
ENDDO
```

The arguments for the MULTIPLY and ADD units must be on a 128-bit boundary to use the SSE or packed instruction. This is a severe restriction; however, the compiler can employ operand shifts to align the operands in the 128 bit registers. Unfortunately, these shift operations will detract from the overall performance of the DO loops. This loop was run with the following two allocations:

```
PARAMETER (IIDIM=100)
COMMON A(IIDIM),AF,B(IIDIM),BF,C(IIDIM) ==> run I
COMMON A(IIDIM),B(IIDIM),C(IIDIM) ==> run II
REAL*8 A,B,C,AF,BF,SCALAR
```

Since IIDIM is even, we know that the arrays will not be aligned in run I since there is a single 8-byte operand between A and B. On run II the arrays are aligned on 128-bit boundaries. The difference in performance follows:

	Run I	Run II
MFLOPS	208	223

This run was made repetitively to make sure that the operands were in cache after the first execution of the loop, and so the difference in these two runs is simply the additional alignment operations. We get a 10% difference in performance when the arrays are aligned on 128-bit boundaries. This is yet another example where memory alignment impacts the performance of an application.

Given these alignment issues, applications which stride through memory and/or use indirect addressing will suffer significantly from alignment problems and usually the compiler will only attempt to use SSE instructions for DO loops that access memory contiguously.

In addition to the floating point ADD and MULTIPLY, the memory functions (LOAD, STORE, MOVE) are also as wide as the floating point units. On the SSE2 64 bit wide instructions, there were 128 bit wide memory functions. On the multicores with the SSE2 instructions, vectorization was still valuable, not because of the width of the floating point units which were only 64 bits wide, but because of the memory operations which could be done as 128-bit operations.

1.3 HARDWARE DESCRIBED IN THIS BOOK

In the first release of the book, we will be concentrating on one of the recent chips from AMD. All the examples in the book have been run on the AMD Magny-Cours with two six-core sockets sharing memory. The competitor to the Magny-Cours is the Intel® Nehalem™ also available with two multicore sockets sharing memory.

The following diagram (Figure 1.8) shows some earlier chips from AMD and Intel, the AMD Barcelona and the Intel Nehalem and a third chip, the Harpertown is presented to illustrate the dynamic change that Intel incorporated in the development of the Nehalem. The Nehalem looks much more like an Opteron™ than any previous Xeon® processor. The diagram

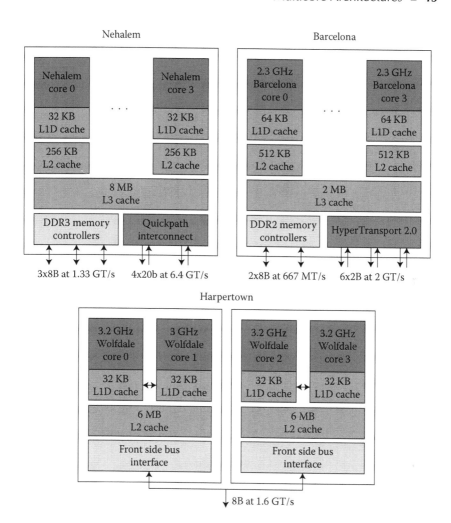

FIGURE 1.8 System architecture comparison, B indicates bytes, b indicates bits. (Adapted from Kanter, D. Inside Nehalem: Intel's future processor and system. http://realworldtech.com/includes/templates/articles.cfm; April 2, 2008.)

shows AMD's Barcelona socket; the differences between the current Magny-Cours and the Barcelona are the size of Level 3 Cache (2 MB on Barcelona and 6 MB on Magny-Cours), the clock cycle, the number of cores per socket and the memory controller. The diagram shows only one of the quad core sockets. Many HPC systems are delivered with two or four of the multicores sharing memory on the node.

Like Opteron's HyperTransport™, the Nehalem introduced a new high-speed interconnect called "QuickPath" which provides a by-pass to the traditional PCI buses. As with HyperTransport on the Barcelona, QuickPath allows higher bandwidth off the node as well as between cores on the node. Both the AMD Magny-Cours and the Nehalem use DDR3 memory and its effective memory bandwidth is significantly higher than the Barcelona. Notice in the diagram that the Nehalem can access three 8-byte quantities at 1.33 GHz, while the Barcelona access two 8-byte quantities at .800 GHz. This is an overall bandwidth improvement of 2.5 in favor of the DDR3 memory systems. When a kernel is memory bound, the performance difference should be greater and when kernel is compute bound, the differences are less.

Note in Figure 1.8 that each core on the socket has its own Level 1 and Level 2 caches and all four cores on the socket share a Level 3 cache. When an application is threaded, either with OpenMP™ and/or Pthreads, data may be shared between the four cores without accessing memory. The mechanism is quite simple if the cores are reading a cache line and it becomes more complicated as soon as a core wants to write into a cache line. While using an OpenMP there is a performance hit referred to as "false sharing" when multiple cores are trying to store into the same cache line. There can only be one owner of a cache line and you must own a cache line when modifying it.

Additionally, the user should understand that the memory bandwidth is a resource that is shared by all the cores on the socket. When a core performs a very memory-intensive operation, say adding two large arrays together, it would require a significant portion of the total memory bandwidth. If two cores are performing the memory-intensive operation, they have to share the available memory bandwidth. Typically, the memory bandwidth is not sufficient to have every core performing a memory-intensive operation at the same time. We will see in Chapter 8 that OpenMP tends to put more pressure on the memory bandwidth, since the execution of each core is more tightly coupled than in an MPI application.

EXERCISES

1. What is the size of TLB on current AMD processors? How much memory can be mapped by the TLB at any one time? What are possible causes and solutions for TLB thrashing?
2. What is cache line associativity? On x86 systems how many rows of associativity are there for Level 1 cache? How can these cause performance problems? What remedies are there for these performances?

3. How does SSE3 improve on the original SSE2 instructions. What can prevent the compiler from using SSE instructions directly?

4. In relation to cache, what is false sharing?

5. What is the size of typical Level 1 cache, Level 2 cache, and Level 3 cache?

6. How is performance affected by memory-intensive operations running simultaneously on multiple cores? Why?

7. What are some of the reasons why a user may choose to use only 1/2 of the cores on a node? On a two socket node, is it important to spread out the cores being used evenly between the sockets?

9. What are some of the improvements in going from the Harperstown chip to the Nehalem?

CHAPTER 2

The MPP

A Combination of Hardware and Software

W HEN AN APPLICATION RUNS across a massively parallel processor (MPP), numerous inefficiencies can arise that may be caused by the structure of the application, interference with other applications running on the MPP system or inefficiencies of the hardware and software that makes up the MPP. The interference due to the other applications can be the source of nonreproducibility of runtimes. For example, all applications run best when they run on an MPP in a dedicated mode. As soon as a second application is introduced, either message passing and/or I/O from the second application could perturb the performance of the first application by competing for the interconnect bandwidth. As an application is scaled to larger and larger processor counts, it becomes more sensitive to this potential interference. The application programmer should understand these issues and know what steps can be taken to minimize the impact of the inefficiencies. The issues we cover in this chapter are

1. Topology of the interconnect and how its knowledge can be used to minimize interference from other applications

2. Interconnect characteristics and how they might impact runtimes

3. Operating system jitter

2.1 TOPOLOGY OF THE INTERCONNECT

The topology of the interconnect dictates how the MPP nodes are connected together. There are numerous interconnect topologies in use today. As MPP systems grow to larger and larger node counts, the cost of some interconnect topologies grows more than others. For example, a complete crossbar switch that connects every processor to every other processor becomes prohibitedly expensive as the number of processors increase. On the other hand, a two- or three-dimensional torus grows linearly with the number of nodes. All very large MPP systems such as IBM®'s Blue Gene® [5] and Cray's XT™ [6] use a 3D torus. In this discussion, we will concentrate on the torus topology.

In the 3D torus, every node has a connection to its north, south, east, west, up, and down neighboring nodes. This sounds ideal for any mesh-based finite difference, finite-element application, or other application that does nearest-neighbor communication. Of course within the node itself, all the cores can share memory and so there is a complete crossbar interconnect within the node. When mapping an application onto an MPP system, the scheduler and the user should try to keep a decomposed neighborhood region within the node. In this way, the MPI tasks within the node can take advantage of the ability to perform message passing using shared memory moves for the MPI tasks within the node. Additionally, it would be good to make sure that the MPI task layout on the interconnect is performed in a way where neighboring MPI tasks are adjacent to each other on the torus. Imagine that the 3D simulation is laid out on the torus in the same way as the physical problem is organized. For example, consider an application that uses a $1000 \times 1000 \times 10$ 3D mesh and that the grid is decomposed in $10 \times 10 \times 5$ cubical sections. There are therefore $100 \times 100 \times 2$ cubical sections. If we want to use 20,000 processors we would want to lay out the cubical sections on the nodes so that we have a $2 \times 2 \times 2$ group of cubical sections on a two-socket quad-core node. This leaves us with a 2D array of 50×50 nodes. This application could therefore be efficiently mapped on a 2D torus of 50×50 nodes and all nearest neighbor communication would be optimized on the node as well as across the interconnect. Such a mapping to the MPP will minimize the latency and maximize the bandwidth for the MPI message passing.

What is good for a single application may not be best for the overall throughput of the MPP system. The previous mapping discussion is the best thing to do when running in dedicated mode where the application has access to the entire system; however, what would happen if there were several large applications running in the system? Consider an analogy. Say we pack eggs in large cubical boxes holding 1000 eggs each in a $10 \times 10 \times 10$ configuration. Now we want to ship these boxes of eggs in a truck that has a dimension of $95 \times 45 \times 45$. With the eggs contained in the $10 \times 10 \times 10$ boxes we could only get 144,000 ($90 \times 40 \times 40$) eggs in the truck. If, on the other hand, we could split up some of the $10 \times 10 \times 10$ boxes into smaller boxes we could put 192,375 ($95 \times 45 \times 45$) eggs in the truck. This egg-packing problem illustrates one of the trade-offs in scheduling several MPI jobs onto a large MPP system. Given a 3D torus as discussed earlier, it has three dimensions which correspond to the dimensions of the truck. There are two techniques that can be used. The first does not split up jobs; it allocates the jobs in the preferred 3D shape and the scheduler will never split up jobs. Alternatively, the scheduler allocates the jobs in the preferred 3D shape until the remaining holes in the torus are not big enough for another contiguous 3D shape and then the scheduler will split up the job into smaller chunks that can fit into the holes remaining in the torus. In the first case, a large majority of the communication is performed within the 3D shape and very few messages will be passed outside of that 3D shape. The second approach will utilize much more of the system's processors; however, individual jobs may interfere with other jobs since their messages would necessarily go through that portion of the torus that is being used for another application.

2.1.1 Job Placement on the Topology

The user cannot change the policy of the computing center that runs the MPP system; however, there are several options the user can take to try to get as much locality as possible. For example, the scheduler does not know what kind of decomposition a particular application might have and therefore it is at a loss as to the shape of a desired 3D or 2D shape to minimize nearest-neighbor communication. To remedy this situation, the mpirun command needs to have an option for the user to specify how to organize the MPI tasks on the topology of the interconnect. There are typically mechanisms to group certain MPI tasks on nodes, to take advantage of the shared memory communication as well as mechanisms to allocate MPI tasks across the nodes in a certain shape. If only one application can have access to a given node, the internode allocation would always be honored by the scheduler; however, when the scheduler is trying to fill up the entire system, the allocation of nodes within a certain shape is just a suggestion and may be overridden by the scheduler. Later we will see how the performance of an application can be improved by effectively mapping the MPI task onto the 3D torus.

2.2 INTERCONNECT CHARACTERISTICS

The interconnect can be designed to improve the performance of disjoint applications that arise from using the second scheduler strategy. For example, if the bandwidth of the torus links is higher than the injection bandwidth of a node, then there will be capacity for the link bandwidth to handle more than the typical nearest-neighbor communication. Additionally, if the latency across the entire MPP system is not much greater than the latency between nearest nodes, then the distance between neighboring MPI tasks may not seem so long. The latency across an interconnect is a function of both hardware and software. In a 3D torus, the time for a message to pass through an intersection of the torus can be very short and the numerous hops to get from one part of the interconnect to another may not greatly increase the latency of a neighbor message that has to go across the entire system.

Other characteristics of the interconnect can impact the latency across the entire machine. In a very large application, the likelihood of link errors can be high. Some interconnects only detect an error at the receiving end of the message and then asks the sender to resend. The more sophisticated interconnects can correct errors on the link which will not significantly impact the overall message time. Additionally, as the number of messages in the system increases with large node counts and global calls such as MPI_ALLTOALL and MPI_ALL_REDUCE, the interconnect may exceed its capability for keeping track of the messages in flight. If the number of active connections exceeds the interconnect's cache of messages, then the amount of time to handle additional messages increases and causes perturbation in the time to send the messages.

2.2.1 The Time to Transfer a Message

The time to transfer a message of N bytes from one processor to another is

```
Time to transfer N bytes = Latency of the interconnect
    + N bytes * bandwidth of the interconnect
```

The latency is a function of the location of the sender and the receiver on the network and how long it takes the operating system at the receiving end to recognize that it has a message. The bandwidth is a function of not only the hardware bandwidth of the interconnect, but also what other traffic occupies the interconnect when the message is being transferred. Obviously, the time to transfer the message would be better if the application runs in the dedicated mode. In actuality, when an application is running in a production environment, it must compete with other applications running on the MPP system. Not only does it compete with message passing from the other applications, but also any I/O that is performed by any of the applications that run at the same time necessarily utilize the interconnect and interfere with MPI messages on the interconnect.

2.2.2 Perturbations Caused by Software

Very few, if any, MPP systems have a systemwide clock. Maintaining a consistent clock across all the nodes in the system would be a nightmare. Why does one need a consistent clock? On all MPP systems, each core runs an operating system that is designed to perform many different functions in addition to running an application. For example, Linux®, a widely used operating system, has the ability to handle numerous services such as, disk I/O, sockets, and so on. Each of these requires a demon running to handle a request and Linux must schedule the demons to service potential requests. The more the services available in the operating system, the more the demons must be run to service different requests. On a small system of 500 or less nodes, the time to perform these services may not impact the performance of an application. As the number of nodes used by a given applications increases, the likelihood that these demons interfere with the application increases. This interference typically comes when an application comes to a barrier. Say, a large parallel application must perform a global sum across all the MPI tasks in the system. Prior to performing the sum, each processor must arrive at the call and have its contribution added to the sum. If some processors are off-servicing operating system demons, the processor may be delayed in arriving at the sum. As the number of processors grows, the time to synchronize prior to the summation can grow significantly, depending on the number of demons the operating system must service.

To address operating system jitter, both IBM's Blue Gene and Cray's XT systems have "light-weight" operating system kernels that reduce the number of demons. Consequently, many of these "light-weight" operating systems cannot support all of the services that are contained in a normal Linux system.

2.3 NETWORK INTERFACE COMPUTER

In addition to the handling of the messages coming off the node, in some interconnects, the network interface computer (NIC) must handle the messages and I/O that pass through it from other nodes, going to other nodes. For nearest-neighbor communication, it is sufficient that the link bandwidth of the interconnect be the same as the injection bandwidth from each of the nodes; however, when global communication and I/O is performed, the NICs must not only handle the local communication, but also route messages from distant nodes to other distant nodes. To handle significant global communication, as in a 3D transpose of an array, the link bandwidth has to handle much more traffic than just the processors running on the nodes to which it is connected. An important capability of the NIC is referred to as adaptive routing. Adaptive routing is the notion of being able to send a message on a different path to the destination depending on the traffic in the network or a disabled NIC. Adaptive routing can significantly improve the effective bandwidth of the network and can be utilized when the MPP system has the ability to identify when an NIC on the network is down and reconfigure the routing to bypass the down NIC.

Another very important characteristic of the NIC is the rate at which it can handle messages. As the number of cores on the multicore node grows, the NIC's ability to handle all the messages that come off the node becomes a bottleneck. As the number of messages exceeds its rate of handling messages, there is a slowdown that may introduce interference into the MPP system.

2.4 MEMORY MANAGEMENT FOR MESSAGES

Frequently, when a message is received by a node, memory must be dynamically allocated to accommodate the message. The management of this message space is a very complex problem for the operating system. On the Linux operating system with small pages of size 4096 bytes, the allocation, deallocation, and garbage collection for the message space can take an inordinate amount of time. Very often, this memory can become fragmented which impacts the memory transfer rates. The previous section discussed the use of small pages and how they may not be contiguous in memory and so the actual memory bandwidth may suffer as a result of the disjointed pages.

The application programmer can help in this area. By preposting the receives prior to sending data, the message may not be held in intermediate message buffers, but may be delivered directly to the application buffer which has a much better chance of being contiguous in memory. More will be discussed on this in the MPI section.

2.5 HOW MULTICORES IMPACT THE PERFORMANCE OF THE INTERCONNECT

As the number of cores on the multicore node increases, more pressure is placed on the interconnect. When an application is using all MPI, the number of messages going out to the interconnect increases as the number of cores increases. It is a common occurrence to move an application from an 8- to a 12-core node and have the overall performance of the application decrease. How can the user deal with the problems that arise as a system is upgraded to larger multicore sockets?

The first thing the user can do is to make sure that the minimum amount of data is shipped off the node. By organizing the optimal neighborhood of MPI tasks on the multicore node, the number of messages going off the node can be minimized. Messages that are on the node can be performed with efficient shared memory transfers.

The second approach would be to introduce OpenMP™ into the all-MPI application, so that fewer MPI tasks are placed on the multicore node. In the next 4–5 years the number of cores on a node will probably increase to 24–32. With such a fat node, the use of MPI across all the cores on the multicore node will cease to be an efficient option. As multicore nodes are designed with larger number of cores, the days of an application using MPI between all of the cores of a MPP system may be nearing the end.

The reluctance of application developers to introduce OpenMP into their applications may disappear once the GPGPUs emerge as viable HPC computational units. Since GPGPUs require some form of high-level shared memory parallelism and low-level vectorization, the same programming paradigm should be used on the multicore host. A popular code design for using attached GPGPUs is to have an application that uses MPI between nodes and OpenMP on the node, where high-level kernels of the OpenMP are computed on the GPGPU. Ideally, an application can be structured to run efficiently on a multicore MPP without GPGPUs or to run efficiently on a multicore MPP with GPGPUs simply with comments line directives. This approach will be discussed in Chapter 9.

EXERCISES

1. The number of MPI messages that an NIC can handle each second is becoming more and more of a bottleneck on multicore

systems. The user can reduct the number of messages sent by assuring the best decomposition on the node. Given an all-MPI application using 8 MPI tasks on the node, how would the number of messages for one halo exchange change in the following three scenarios?

   a. The MPI tasks on the node are not neighbors of other MPI tasks on the node.
   b. The MPI tasks on the node are arranged in a $2 \times 4 \times 1$ plane of a grid.
   c. The MPI tasks on the node are arranged in a $2 \times 2 \times 2$ cube.

2. Having seen that the decomposition can impact halo exchange performance, can it also impact the global sum of a single variable? How about the performance of a 3D ALL to ALL?

3. When an application is memory bandwidth limited users may be tempted to use fewer cores/node and more nodes. If the computer center charges for Node/hour will this method ever result in reduced cost? What if the computer center charges for power used?

4. What are some typical interconnect topologies? Why would the manufacturer choose one over another? What might be the advantages and disadvantages of the following interconnects?

   a. Full all-to-all switch
   b. Fat Tree
   c. Hypercube
   d. 2D Torus
   e. 3D Torus

5. Why would an application's I/O impact other users in the system?

6. What is operating system jitter? When and why does it interfere with performance? How is it minimized?

7. What is injection bandwidth? How can communication from one job interfere with the performance of a different job?

8. Given an interconnect with an injection bandwidth of 2 GB/s and a latency of 5 μs, what is the minimum time to send the following messages:

   a. 100 messages of 80 bytes
   b. 10 messages of 800 bytes
   c. 1 message of 8000 bytes

9. As multicore nodes grow larger and larger, how might this impact the design of the interconnect?

   a. In injection bandwidth?
   b. In message/s handled?
   c. In the topology of the interconnect?

CHAPTER 3

How Compilers Optimize Programs

To MANY APPLICATION DEVELOPERS, compilation of their code is a "black art." Often, a programmer questions why the compiler cannot "automatically" generate the most optimal code from their application. This chapter reveals the problems a compiler faces generating an efficient machine code from the source code, and gives some hints about how to write a code to be more amenable to automatic optimization.

3.1 MEMORY ALLOCATION

The compiler has several different ways of allocating the data referenced in a program. First, there is the notion of static arrays; that is, arrays that are allocated at compile time. Whenever all the characteristics of a variable are known, it may be allocated in the executable and therefore the data are actually allocated when the executable is loaded. Most modern applications dynamically allocate their arrays after reading the input data and only then allocate the amount of data required. The most often used syntax to do this is the typical malloc in C and/or using ALLOCATABLE arrays in Fortran. A more dynamic way to allocate data is by using automatic arrays. This is achieved in Fortran by passing a subroutine integers that are then used to dynamically allocate the array on the subroutine stack.

The efficiency of an application heavily depends upon the way arrays are allocated. When an array is allocated within a larger allocation, and that allocation is performed only once, the likelihood that the array is allocated on contiguous physical pages is very high. When arrays are frequently allocated and deallocated, the likelihood that the array is allocated on contiguous physical pages is very low. There is a significant benefit to having an array allocated on contiguous physical pages. Compilers can only do so much when the application dynamically allocates and deallocates memory. When all the major work arrays are allocated together on subsequent ALLOCATE statements, the compiler usually allocates a large chunk of memory and suballocates the individual arrays; this is good. When users write their own memory allocation routine and call it from numerous locations within the application, the compiler cannot help to allocate the data in contiguous pages. If the user must do this it is far better to allocate a large chunk of memory first and then manage that data by cutting up the memory into smaller chunks, reusing the memory when the data is no longer needed. De-allocation and re-allocating memory is very detrimental to an efficient program. Allocating and deallocating arrays end up in increasing the compute time to perform garbage collection. "Garbage collection" is the term used to describe the process of releasing unnecessary memory areas and combining them into available memory for future allocations. Additionally, the allocation and deallocation of arrays leads to memory fragmentation.

3.2 MEMORY ALIGNMENT

As discussed in Chapter 2, the way the program memory is allocated impacts the runtime performance. How are arrays, which are often used together, aligned, and does that alignment facilitate the use of SSE instructions, the cache, and the TLB? The compiler tries to align arrays to utilize cache effectively; however, the semantics of the Fortran language do not always allow the compiler to move the location of one array relative to another. The following Fortran constructs inhibit a compiler from padding and/or aligning arrays:

1. Fortran COMMON BLOCK

2. Fortran MODULE

3. Fortran EQUIVALENCE

4. Passing arrays as arguments to a subroutine

5. Any usage of POINTER

6. ALLOCATABLE ARRAYS

When the application contains any of these structures, there can be an implicit understanding of the location of one array in relation to another. Fortran has very strict storage and sequence association rules which must be obeyed when compiling the code. While COMMON blocks are being replaced by MODULES in later versions of Fortran, application developers have come to expect the strict memory alignment imposed by COMMON blocks. For example, when performing I/O on a set of arrays, application developers frequently pack the arrays into a contiguous chunk of logical memory and then write out the entire chunk with a single write. For example,

```
COMMON A(100,100), B(100,100), C(100,100)
WRITE (10) (A(I), I=1, 30000)
```

Using this technique results in a single I/O operation, and in this case, a write outputting all A, B, and C arrays in one large block, which is a good strategy for efficient I/O. If an application employs this type of coding, the compiler cannot perform padding on any of the arrays A, B, or C.

Consider the following call to a subroutine crunch.

```
CALL CRUNCH (A(1,10), B(1,1), C(5,10))
ooo
SUBROUTINE CRUNCH (D,E,F)
DIMENSION D(100), E(10000), F(1000)
```

This is legal Fortran and it certainly keeps the compiler from moving A, B, and C around. Unfortunately, since Fortran does not prohibit this type of coding, the alignment and padding of arrays by compilers is very limited.

A compiler can pad and modify the alignment of arrays when they are allocated as automatic or local data. In this case the compiler allocates memory and adds padding if necessary to properly align the array for efficient access. Unfortunately, the inhibitors to alignment and padding far outnumber these cases. The application developer should accept responsibility for allocating their arrays in such a way that the alignment is conducive to effective memory utilization. This will be covered in more in Chapter 6.

3.3 VECTORIZATION

Vectorization is an ancient art, developed over 40 years ago for real vector machines. While the SSE instructions require the compiler to vectorize the Fortran DO loop to generate the instructions, today SSE instructions are not as powerful as the vector instructions of the past. We see this changing with the next generation of multicore architectures. In an attempt to generate more floating point operations/clock cycle, the SSE instructions are becoming wider and will benefit more from vectorization. While some of today's compilers existed and generated a vector code for the past vector processors, they have had to dummy down their analysis for the SSE instructions. While performance increases from vectorization on the past vector systems ranged from 5 to 20, the SSE instructions at most achieve a factor of 2 for 64-bit arithmetic and a factor of 4 for 32-bit arithmetic. For this reason, many loops that were parallelizable with some compiler restructuring (e.g., IF statements) are not vectorized for the SSE instructions, because the overhead of performing the vectorization is not justified by the meager performance gain with the SSE instructions.

With the advent of the GPGPUs for HPC and the wider SSE instructions, using all the complicated restructuring to achieve vectorization will once again have a big payoff. For this reason, we will include many vectorization techniques in the later chapters that may not obtain a performance gain with today's SSE instructions; however, they absolutely achieve good performance gain when moving to a GPGPU. The difference is that the vector performance of the accelerator is 20–30 times faster than the scalar performance of the processing unit driving the accelerator. The remainder of this section concentrates on compiling for cores with SSE instructions.

The first requirement for vectorizing SSE instructions is to make sure that the DO loops access the arrays contiguously. While there are cases when a compiler vectorizes part of a loop that accesses the arrays with a stride and/or indirect addressing, the compiler must perform some overhead prior to issuing the SSE instructions. Each SSE instruction issued must operate on a 128-bit register that contains two 64-bit operands or four 32-bit operands. When arrays are accessed with a stride or with indirect addressing, the compiler must fetch the operands to cache, and then pack the 128-bit registers element by element prior to issuing the floating point operation. This overhead in packing the operands and subsequently unpacking and storing the results back into memory introduces an overhead that is not required when the scalar, non-SSE instructions are issued. That overhead degrades from the factor of 2 in 64-bit mode or 4 in 32-bit mode. In Chapter 5, we examine cases where packing and unpacking of operands for subsequent vectorization does not pay off. Since compilers tend to be conservative, most do not vectorize any DO loop with noncontiguous array accesses. Additionally, when a compiler is analyzing DO loops containing IF conditions, it does not vectorize the DO loop due to the overhead required to achieve that vectorization. With the current generation of multicore chips, application developers should expect that only contiguous DO loops without IFs will be analyzed for vectorization for the SSE instructions.

3.3.1 Dependency Analysis

The hardest part in vectorizing a DO loop is determining if the operations in the loop are independent. Can the loop operations be distributed across all the iterations of the loop? If the loop has a "loop-carried" dependency, the loop cannot be vectorized. Consider this classic example:

```
DO I = N1,N2
    A(I) = A(I+K) + scalar * B(I)
ENDDO
```

If K is −1, the loop cannot be vectorized, but if K is +1, it can be vectorized.

So what happens if K is equal to −1:

- Iteration I = N1: $A(N1) = A(N1 - 1) + scalar * B(N1)$

- Iteration I = N + 1: $A(N1 + 1) = A(N1) + scalar * B(N1 + 1)$

- Iteration I = N1 + 2: $A(N1 + 2) = A(N1 + 1) + scalar * B(N1 + 2)$

(Bold, italic indicates an array element computed in the DO loop.)

Here the element of the A array needed as input to the second iteration is calculated in the first iteration of the loop. Each update after the first requires a value calculated the previous time through the loop. We say that this loop has a loop-carried dependency.

When a loop is vectorized, all the data on the right-hand side of the equal sign must be available before the loop executes or be calculated earlier in the loop, and this is not the case with K = −1 above.

Now, what if K is equal to +1:

- Iteration I = N1: $A(N1) = A(N1 + 1) + scalar * B(N1)$

- Iteration I = N1 + 1: $A(N1 + 1) = A(N1 + 2) + scalar * B(N1 + 1)$

- Iteration I = N1 + 2: $A(N1 + 2) = A(N1 + 3) + scalar * B(N1 + 2)$

Here, all the values needed as input are available before the loop executes. Although $A(N1 + 1)$ is calculated in the second pass of the DO, its old value is used in the first pass of the DO loop and that value is available. Another way of looking at this loop is to look at the array assignment:

```
A(N1 : N2) = A(N1 + K:N2 + K) + scalar * B(N1:N2)
```

Regardless of what value K takes on, the array assignment specifies that all values on the right hand side of the replacement sign are values that exist prior to executing the array assignment. For this example, the array assignment when $K = -1$ is not equivalent to the DO loop when $K = -1$. When $K = +1$, the array assignment and the DO loop are equivalent.

The compiler cannot vectorize the DO loop without knowing the value of K. Some compilers may compile both a scalar and vector version of the loop and perform a runtime check on K to choose which loop to execute. But this adds overhead that might even cancel any potential speedup that could be gained from vectorization, especially if the loop involves more than one value that needs to be checked. Another solution is to have a comment line directive such as

```
!DIR$ IVDEP
```

This directive was introduced by Cray® Research in 1976 to address situations where the compiler needed additional information about the DO loop. A problem would arise if this directive was placed on the DO loop and there were truely data dependencies. Wrong answers would be generated.

When a compiler is analyzing a DO loop in C, additional complications can hinder its optimization. For example, when using C pointers

```
for (i = 0; i < 100; i++) p1[i] = p2[i];
```

p1 and p2 can point anywhere and the compiler is restricted from vectorizing any computation that uses pointers. There are compiler switches that can override such concerns for a compilation unit.

3.3.2 Vectorization of IF Statements

The legacy vector computers had special hardware to handle the vectorization of conditional blocks of the code within loops controlled by IF statements. Lacking this hardware and any real performance boost from vectorization, the effect of vectorizing DO loops containing a conditional code seems marginal on today's SSE instructions; however, future, wider SSE instructions and GPUs can benefit from such vectorization.

There are two ways to vectorize a DO loop with an IF statement. One is to generate a code that computes all values of the loop index and then only store results when the IF condition is true. This is called a *controlled store*. For example,

```
DO I = 1,N
        IF(C(I).GE.0.0)B(I) = SQRT(A(I))
ENDDO
```

The controlled store approach would compute all the values for I = 1, 2, 3, ..., N and then only store the values where the condition C(I).GE.0.0 where true. If C(I) is never true, this will give extremely poor performance. If, on the other hand, a majority of the conditions are true, the benefit can be significant. Control stored treatment of IF statements has a problem in that the condition could be hiding a singularity. For example,

```
DO I = 1,N
        IF(A(I).GE.0.0)B(I) = SQRT(A(I))
ENDDO
```

Here, the SQRT(A(I)) is not defined when A(I) is less than zero. Most smart compilers handle this by artificially replacing A(I) with 1.0 whenever A(I) is less than zero and take the SQRT of the resultant operand as follows:

```
DO I = 1,N
        IF(A(I).LT.0.0) TEMP(I) = 1.0
        IF(A(I).GE.0.0) TEMP(I) = A(I)
        IF(A(I).GE.0.0)B(I) = SQRT(TEMP(I))
ENDDO
```

A second way is to compile a code that *gathers* all the operands for the cases when the IF condition is true, then perform the computation for the "true" path, and finally *scatter* the results out into the result arrays. Considering the previous DO loop,

```
DO I = 1,N
        IF(A(I).GE.0.0)B(I) = SQRT(A(I))
ENDDO
```

The compiler effectively performs the following operations:

```
II = 1
DO I = 1,N
        IF(A(I).GE.0.0)THEN
```

```
            TEMP(II) = A(II)
            II = II + 1
            ENDIF
      ENDDO
      DO I = 1,II-1
            TEMP(I) = SQRT(TEMP(I))
      ENDDO
      II = 1
      DO I = 1,N
            IF(A(I).GE.0.0)THEN
            B(I) = TEMP(II)
            II = II + 1
            ENDIF
      ENDDO
```

This is very ugly indeed. The first DO loop gathers the operands that are needed to perform the SQRT; this is called the gather loop. The second DO loop performs the operation on only those operands that are needed: no problem with singularities. The third DO loop scatters the results back into the B array.

The conditional store method has the overhead of the unnecessary operations performed when the condition is false. The gather/scatter approach avoids unnecessary computation, but it introduces data motion. Because of the overhead these methods introduce, most compilers would not try to vectorize loops containing conditional code. But, there are cases when it can be beneficial. We will look at this in more detail in Chapter 6.

3.3.3 Vectorization of Indirect Addressing and Strides

Since the compiler does introduce register-packing instructions to handle unaligned contiguous array reference, one would think that they should be able to perform similar operations to handle the vectorization of DO loops that contain indirect addressing and strides; however, the compiler cannot determine the alignment of variables within an array when indirect addressing is used. The relative location between array elements within an indirectly addressed array could range from zero to very distant in memory. For example, consider a loop that uses indirect addressing to obtain material properties of variable.

```
DO I = 1, NX
      X(I) = Y(I)*density(mat(I))
ENDDO
```

For some ranges of I, the value of mat(I) may be constant. There is no good reason for the compiler not to vectorize this construct given that knowledge. If, on the other hand, the data being referenced by the indirect address are in completely different cache lines or pages, then the performance gained from vectorizing the loop may be poor.

3.3.4 Nested DO Loops

Nested DO loops appear in most significant computational-intensive codes. How does the compiler decide which loop in the nest to vectorize, and does it have the right information to make this decision?

The compiler has to consider four facts:

1. Which loop accesses most of the arrays contiguously?

2. Which loop has the most iterations?

3. Which loop can be vectorized?

4. What ordering of the loops would benefit cache reuse the most?

The loop that accesses the arrays contiguously is the easiest to identify and the most important to get right. There are situations when the loop accessing the arrays contiguously still might not be the best loop to have as an inner loop. For example, if that loop's iteration count is very small, say 1 to 4, the gain from accessing arrays contiguously may not be as great as choosing a longer loop, or even eliminating the short contiguous loop by unrolling it.

Determining which of the loops is the longest (in iteration count) might not be easy. If the arrays and their sizes are subroutine parameters, the compiler might not be able to determine which DO loop is going to have the longest iteration count at runtime. Even in interprocedural analysis, where the compiler uses precompiled information about other routines, the routine could be called from more than one location with different size arrays, like a matrix multiply routine called from many places in the code and with very different array sizes in each call. Once again, the compiler has its hands tied from doing the smart thing. Later, we will examine numerous examples where the user can assist in generating the most efficient code.

There are many ways to reuse elements in a cache line that is resident in cache (cache reuse). The more reuse, the fewer memory fetches and the better the performance. *Cache blocking* or *tiling* is one way of achieving efficient cache reuse. Other techniques tend to be more difficult for the compiler to perform, and involve code restructuring, which we will discuss in Chapter 6.

With cache blocking, the compiler breaks a nested DO loop into computational blocks that work on data that can reside in cache during the execution of the block. Cache blocking does not benefit all nested DO loops, because there are conditions that must be met for cache blocking to achieve a performance gain.

3.4 PREFETCHING OPERANDS

We discussed prefetching in an earlier section. Prefetching upcoming operands to overlap the computation of one iteration of a loop while the fetching of operands for the next iteration is a good optimization strategy since the latency of fetching operands to the cache is so much longer than fetching operands from memory.

Getting the right amount of prefetching is difficult, since the operands must be fetched to the same caches that are being used to hold the data that is currently being processed. In some cases, prefetching should be turned off when the programmer has restructured the code to do its own cache blocking and alignment.

Some compilers give informative messages regarding any prefetching. For example, in the following DO loop,

```
(69)
(70)      DO 46021 I = 1,  N
(71)      A(I,1) = B(I,1) * C(1,1) + B(I,2) * C(2,1)
(72)      *        + B(I,3) * C(3,1) + B(I,4) * C(4,1)
(73)      A(I,2) = B(I,1) * C(1,2) + B(I,2) * C(2,2)
(74)      *        + B(I,3) * C(3,2) + B(I,4) * C(4,2)
(75)      A(I,3) = B(I,1) * C(1,3) + B(I,2) * C(2,3)
(76)      *        + B(I,3) * C(3,3) + B(I,4) * C(4,3)
(77)      A(I,4) = B(I,1) * C(1,4) + B(I,2) * C(2,4)
(78)      *        + B(I,3) * C(3,4) + B(I,4) * C(4,4)
(79)  46021 CONTINUE
(80)
```

The Portland Group compiler generated the following message:

70, Generated an alternate loop for the inner loop

Generated vector SSE code for inner loop

Generated four prefetch instructions for this loop

Generated vector SSE code for inner loop

Generated four prefetch instructions for this loop

The first message indicates that an alternate loop was generated for the inner loop to handle the case when the arrays are not aligned. Vector SSE instructions were generated for the loop as well as four prefetch instructions. The four lines being prefetched are probably the next cache lines of B(I,1), B(I,2), B(I,3), and B(I,4).

3.5 LOOP UNROLLING

Another powerful restructuring technique used by many compilers is loop-unrolling. When the compiler knows the value of the loop index and the loop index is constant, it may choose to completely unroll the DO loop and get rid of the loop entirely. Consider the following DO loop:

```
(45)    DO 46020 I = 1,N
(46)    DO 46020 J = 1,4
(47)    A(I,J) = 0.
(48)    DO 46020 K = 1,4
(49)    A(I,J) = A(I,J) + B(I,K) * C(K,J)
(50)    46020 CONTINUE
```

The Portland Group compiler gives the following messages:

45, Generated an alternate loop for the inner loop

Generated vector SSE code for inner loop

Generated four prefetch instructions for this loop

Generated vector SSE code for inner loop

Generated four prefetch instructions for this loop

46, Loop unrolled four times (completely unrolled)

48, Loop not vectorized: loop count too small

Loop unrolled four times (completely unrolled)

Here we have a case where the compiler did the unrolling and then did the vectorization and prefetching. This is an extremely powerful optimization for this loop nest which results in factors of 4–5 in performance gain.

3.6 INTERPROCEDURAL ANALYSIS

If only the compiler knew more about the subroutine it is analyzing, it could do a better job generating an optimal code. Even though Fortran 90 introduced INTERFACE blocks into the language for the user to provide this useful information to the compiler, since it is not required, few programmers use the syntax, leaving the compiler to do its own analysis. *Interprocedural analysis* requires the compiler to look at a large portion of the application as a single compilation unit. At the basic level, the compiler needs to retain information about the arguments that are passed from the caller to the called routine. The compiler can use that information to perform deeper optimizations of the called routine, especially when literal constants are passed for a variable in a call. The most drastic interprocedural optimization is to completely inline all the called routines that can be inlined. While this could increase the compile time, significant performance gains frequently can be achieved by eliminating costly subprogram call/returns.

For example, PGI's compiler, like many others, has an option for performing interprocedural analysis (IPA = inline, fast) that generates additional data files during compilation that are used for optimization analysis. In one particular case, the compile time increased from 4–5 min to 20–30 min and the performance gain was 10–15% of the computation time. If this application is infrequently compiled and the executable is used for thousands of hours of computation, this is a good trade-off.

3.7 COMPILER SWITCHES

All the compilers' optimization options can be controlled through the use of command line option flags and comment line directives. When fine tuning an application, the programmer can use the compile line option across the entire code or use comment line directives on selected portions of the code. There are options that control every compiler optimization function. Each compiler has its own set of options, but they all provide control over the level of optimization the compiler would use. It is important to familiarize yourself with the options available with the compiler you are using. Some that are very important are

1. Vectorization is typically not done by default; vectorization should always be tried.

2. Interprocedural analysis is not always performed. This is very important for applications that call numerous small routines and in particular for C and C++ applications.

3. Unrolling, cache blocking, prefetching, and so on. These optimizations are usually combined into a general optimization flag like –fast; however, many cases may benefit by selectively utilizing comment line directives on specific DO loops.

4. Always get information about what optimization the compiler performed. The PGI compiler has both a –Minfo flag and a –Mneginfo flag, the first for the compiler to show what optimizations took place in the compilation of a routine and the second to give reasons why some optimizations, such as vectorization, could not be performed.

5. Do not use automatic shared memory parallelization. Automatic shared memory parallelization does not give the best OpenMP™ performance. In some cases, one part of an application might get a good speedup while another portion gives a slowdown. Blindly using automatic parallelization is not recommended, unless runtime statistics are gathered to show which routines are improved and which are degraded.

Appendix A has a table of important compiler switches for the PGI compiler and the Cray compilation environment.

3.8 FORTRAN 2003 AND ITS INEFFICIENCIES

With the development of Fortran 90, 95, and now 2003, new semantics have been introduced into the language, which are difficult or even impossible to compile efficiently. As a result, programmers are frequently disappointed with the poor performance when these features are used.

Here are a few of the newer Fortran features that would cause most compilers to generate an inefficient code and should be avoided:

1. Array syntax

2. Calling standard Fortran functions not linked to optimized libraries

3. Passing array sections

4. Using modules for local data

5. Derived types

3.8.1 Array Syntax

Array syntax was first designed for the legacy memory-to-memory vector processors such as the Star 100 from Control Data Corporation. The intent of the syntax was to give the compiler a form they could convert directly into a memory-to-memory vector operation. When these vector machines retired, array syntax was kept alive by Thinking Machine's *Connection Machine*. Here the compiler could generate an SIMD parallel machine instruction that would be executed by all the processors in a lock-step parallel fashion.

Unfortunately, array syntax is still with us and while many programmers feel it is easier to use than the standard old Fortran DO loop, the real issue is that most Fortran compilers cannot generate good cache efficient code from a series of array assignment statements.

Consider the following sequence of array syntax from the Parallel Ocean Program (POP) [7]. In this example, all the variables that are in capitalized letters are arrays. In the test that we run, the sizes of the arrays are (500,500,40). Each array assignment completely overflows the TLB and all levels of cache and when the next array assignment is performed, the data will have to be reloaded from the higher level caches and/or memory.

```
!
! DP_1/DT
      WORK3 = mwjfnums0t1 + TQ * (c2 * mwjfnums0t2 +           &
         c3 * mwjfnums0t3 * TQ) + mwjfnums1t1 * SQ
    ! DP_2/DT
      WORK4 = mwjfdens0t1 + SQ * mwjfdens1t1 +                 &
            TQ * (c2* (mwjfdens0t2 + SQ * SQR * mwjfdensqt2)+  &
            TQ * (c3* (mwjfdens0t3 + SQ * mwjfdens1t3) +       &
            TQ * c4*mwjfdens0t4))
      DRHODT = (WORK3 - WORK1 * DENOMK * WORK4) * DENOMK
```

Most of the variables shown are multidimensioned arrays. Today's compilers generate three looping structures around each of the statements. When the arrays used in the first array assignment are larger than what can be fit in Level 1 cache, those variables used in both the first and second array assignment have to be fetched from Level 2 cache and if the arrays are larger than what can be held in Level 2 cache, they have to be retrieved from Level 3 cache or memory. On the other hand, writing the three statements in a single looping structure results in better cache utilization.

```
! DP_1/DT
DO K = 1, NZBLOCK
  DO J = 1, NYBLOCK
    DO I = 1, NXBLOCK
      WORK3 (I, J, K) = mwjfnums0t1 + TQ (I, J, K) * (c2 * mwjfnums0t2 + &
        c3 * mwjfnums0t3 * TQ (I, J, K)) + mwjfnums1t1 * SQ (I, J, K)
      WORK4 (I, J, K) = mwjfdens0t1 + SQ (I, J, K) * mwjfdens1t1 +      &
        TQ (I, J, K) * (c2 * (mwjfdens0t2 + SQ (I, J, K) * SQR (I, J, K)
             * mwjfdensqt2) +                                          &
        TQ (I, J, K) * (c3 * (mwjfdens0t3 + SQ (I, J, K) * mwjfdens1t3) + &
        TQ (I, J, K) * c4 * mwjfdens0t4))
      DRHODT (I, J, K) = (WORK3 (I, J, K) - WORK1 (I, J, K) *
             DENOMK (I, J, K) *                                        &
        WORK4 (I, J, K)) * DENOMK (I, J, K)
    ENDDO
  ENDDO
ENDDO
```

Now the compiler generates a very efficient cache-friendly code. The variables used in the three statements are only fetched once from memory and subsequent uses come from Level 1 cache. Each sweep through the I loop fetch up cache lines which can then be reused in subsequent statements in the DO loop. Additionally, we get much better utilization of the TLB since we are uniformly accessing each array. We get the following impressive speedup:

	Time (s)	TLB Refs/Miss	Level 1 Cache Hits
Original array syntax	0.240	398	95.8%
Fortran DO loop	0.138	533	98.3%

The POP code is an extremely well-written code and a majority of the major computational kernels are written in DO loops. However, this sample code from the equation of state is written in Fortran 90 array syntax and it can definitely be improved. Interestingly, when the number of processors are increased for a given grid size, the size of the arrays on each processor becomes smaller. At some point, the restructuring shown above would not be needed since the size of all the arrays would fit in Level 2 cache. At this point, POP exhibits superlinear scaling. The superlinear scaling comes about, because the arrays assignments work on arrays that fit in Level 1 and Level 2 cache. When control passes from one array assignment to the next, the arrays are still in cache and the restructuring shown above would be unnecessary.

3.8.2 Using Optimized Libraries

This is simply a matter of understanding what the compiler does with calls such as MATMUL and DOT_PRODUCT. Whenever an application uses the more compute intensive intrinsic, make sure that the compiler uses the appropriate routine from the optimized libraries that the vendor supplies. The safest approach is to call the BLAS and LAPACK routines directly rather than trust the Fortran 2000 intrinsics. Some compilers have command line options to automatically recognize standard library calls and link them to the appropriate optimized versions.

3.8.3 Passing Array Sections

Whenever you find a subroutine call that is using a significant amount of time, it is probably because an array section is being passed to the called subroutine. Consider the following code from the combustion code S3D:

```
DO n = 1,n_spec
   call computeScalarGradient(yspecies(:,:,:,n),
            grad_Ys(:,:,:,n,:))
END DO
```

This call invokes a massive amount of memory movement to copy grad_Ys into a contiguous temporary array to pass to computeScalarGradient. Most compilers notice that yspecies is already a contiguous section of memory and they do not perform any data movement on it. The grad_Ys array is copied to a completely separate portion of memory that is passed to the routine. Then on return from the routine the result is copied back into grad_Ys. Typically, this copy is performed with an optimized memory copy routine. If you see an unusual amount of time being spent in a system copy, it could be due to this sort of array section passing. This is a good example of where an Fortran 90 INTERFACE block might help. If grad_Ys is only used as an input to the routine, then the memory operation of moving grad_Ys back to its original form is not required. If, on the other hand, the routine only uses grad_Ys as an output array, the copy in is not needed.

The best method of passing arrays to a routine is by passing an address. If the subroutine argument were grad_Ys(1,1,1,n,1), that would be an address and no data motion would be performed. But that is not the same as what was used in the original call. The programmer would have to reorganize grad_Ys into a structure where n is on the outermost subscript to achieve the improvement. The programmer would have to reorganize grad_Ys in computeScalarGradient to the same form as the caller or reorganize grad_Ys to have n as the last index in the array.

3.8.4 Using Modules for Local Variables

When OpenMP directives are used at a high level in the call chain, care must be taken to assure the variables used in MODULE are shared. This is due to the inability of the compiler to generate thread private copies of MODULE data. Prior to MODULES, COMMON blocks had the same problem and a special kind of COMMON block (TASKCOMMON) was included in the OpenMP standard to handle this situation.

3.8.5 Derived Types

Derived types can have a dramatic impact on efficient memory usage. Consider the following code extracted from one of the SPEC_OMP benchmarks:

```
!$OMP PARALLEL DO DEFAULT(SHARED) PRIVATE(N)
  DO N=1,NUMRT
    MOTION(N)%Ax=NODE(N)%Minv * (FORCE(N)%Xext-FORCE(N)%Xint)
    MOTION(N)%Ay=NODE(N)%Minv * (FORCE(N)%Yext-FORCE(N)%Yint)
    MOTION(N)%Az=NODE(N)%Minv * (FORCE(N)%Zext-FORCE(N)%Zint)
  ENDDO
!$OMP END PARALLEL DO
```

Following is a summary of hardware counters for this loop:

```
USER / solve_.LOOP@li.329
-------------------------------------------------------------------------
Time%                                        4.5%
Time                                         12.197115 secs
Imb.Time                                     0.092292 secs
Imb.Time%                                    1.0%
Calls                       42.9/sec         523.0 calls
PAPI_L1_DCM                 13.700 M/sec     167144470 misses
PAPI_TLB_DM                  0.448 M/sec     5460907 misses
PAPI_L1_DCA                 89.596 M/sec     1093124368 refs
PAPI_FP_OPS                 52.777 M/sec     643917600 ops
User time (approx)          12.201 secs      32941756956
                                             cycles
                                             100.0%Time
```

```
Average Time per Call      0.023321 sec
CrayPat Overhead:Time                       0.0%
HW FP Ops/User time        52.777 M/sec     643917600 ops
                                            0.5%peak(DP)
HW FP Ops/WCT              52.777 M/sec
Computational intensity    0.02 ops/cycle  0.59 ops/ref
MFLOPS (aggregate)         52.78 M/sec
TLB utilization            200.17 refs/miss  0.391 avg uses
D1 cache hit,miss          84.7% hits        15.3% misses
ratios
D1 cache utilization (M)   6.54 refs/miss    0.817 avg uses
```

Notice the poor TLB utilization; any TLB reference/miss below 512 is not good. While the DO loop appears to reference contiguous memory, since we are accessing elements of a derived type, there is a stride of the number of elements within the derived type.

```
TYPE :: motion_type
  REAL(KIND(0D0))  Px    ! Initial x-position
  REAL(KIND(0D0))  Py    ! Initial y-position
  REAL(KIND(0D0))  Pz    ! Initial z-position
  REAL(KIND(0D0))  Ux    ! X displacement
  REAL(KIND(0D0))  Uy    ! Y displacement
  REAL(KIND(0D0))  Uz    ! Z displacement
  REAL(KIND(0D0))  Vx    ! X velocity
  REAL(KIND(0D0))  Vy    ! Y velocity
  REAL(KIND(0D0))  Vz    ! Z velocity
  REAL(KIND(0D0))  Ax    ! X acceleration
  REAL(KIND(0D0))  Ay    ! Y acceleration
  REAL(KIND(0D0))  Az    ! Z acceleration
END TYPE
TYPE (motion_type), DIMENSION(:), ALLOCATABLE :: MOTION
```

Rather than the arrays being dimensioned within the derived type, the derived type is dimensioned. This results in each of the arrays having a stride of 12, which is hurting both TLB and cache utilization.

3.9 SCALAR OPTIMIZATIONS PERFORMED BY THE COMPILER

3.9.1 Strength Reduction

Compilers can often replace an explicit use of an operation with a less expensive iterative operation in a loop. In particular, a multiplication can be replaced with an addition, and an exponentiation can be replaced with a multiplication. In the following example, an exponentiation is replaced with a multiplication:

Original Exponentiation	Compiler Optimized Equivalent
| ```
DO I = 1,10
   A(I) = X**I
END DO
``` | ```
XTEMP = X
DO I = 1,10
   A(I) = XTEMP
   XTEMP = XTEMP*X
END DO
``` |

And similarly, the compiler can replace a multiply with an add:

Original Multiplication	Compiler Optimized Equivalent
| ```
DO I = 1,10
   A(I) = X*I
END DO
``` | ```
XTEMP = X
DO I = 1,10
   A(I) = XTEMP
   XTEMP = XTEMP + X
END DO
``` |

Strength reduction can also be used to optimize array indexing within a loop, as shown in the next example:

Original code:

```
DIMENSION A(100,10)
DO I = 1,10
   A(3,I) = 0.0
END DO
```

The compiler generates code to calculate the address of each element of the array A. The following pseudocode demonstrates the straightforward calculation of the memory address, taking into account that Fortran indexing begins with 1 and assuming each element of A is 8 bytes:

Pseudocode unoptimized:

```
DO I = 1,10
   address = addr(A) + (3-1) * 8 + (I-1) * 100 * 8
   memory(address) = 0.0
END DO
```

The expression (I-1)*100*8 increases by 800 each time through the loop. The compiler can use this to eliminate the multiplies and optimize the address calculation as in the following:

Pseudocode optimized:

```
address = addr(A) + (3-1) * 8 + 1 * 100 * 8
DO I = 1,10
    memory(address) = 0.0
    address = address + 800
END DO
```

The compiler does the math at compile time and generates a simple set of the starting address:

```
address = addr(A) + 816
```

This optimization is very important, since the majority of time spent in most programs is in loops containing repeated accesses of multidimensional arrays. Array dimensions are often variables rather than constants. The following example is more typical:

Original	Brute Force Translation	Optimized
`DIMENSION A (N1,N2,N3)`	`DIMENSION A (N1,N2,N3)`	`DIMENSION A (N1,N2,N3)`
`DO I = 1,N2`	`DO I = 1,N2`	`address = addr(A) ¥`
`  A(M1,I,M3) = 0.0`	`  address = addr(A) ¥`	`  + 8*((M1 + N1*`
`END DO`	`  + (M1-1)*8 + (I-1)*8*N1`	`  (I + N2*M3)) ¥`
	`  + (M3-1)*8*N1*N2`	`  - (1 + N1*1 + N2))`
	`  memory(address) = 0.0`	`DO I = 1,N2`
	`END DO`	`  memory(address) = 0.0`
		`  address = address + N1`
		`END DO`

The code generated within the loop, where most of the time is spent, is much simpler and faster than the brute force calculation.

3.9.2 Avoiding Floating Point Exponents

The following lines of code are nearly identical:

```
X = A**3
Y = A**3.0
```

The compiler translates the first line into

```
X = A*A*A
```

and translates the second line into

```
Y = EXP(LOG(A)*3.0)
```

Clearly, the second form would execute much slower than the first; hence, a simple rule to follow is to use integer exponents whenever possible.

3.9.3 Common Subexpression Elimination

Compilers try to avoid calculating the same expression unnecessarily over and over again when it can generate code to evaluate it just once, save it, and reuse it as needed. It can do this for expressions where none of the input values change from one use to another. Often the single result can be left in a register, saving additional access time. The compiler has to assume that the order and grouping (commutativity and associativity) of the sub-expressions can be changed without altering the result. (This is not always true with finite precision floating-point arithmetic.)

For example,

```
X = A*B*C*D
Y = E*B*F*D
```

might be compiled as

```
T = B*D
X = A*C*T
Y = E*F*T
```

If T is saved in a register, this optimization saves one multiply and two memory accesses.

Of course, if B or D are modified between the set of X and Y, this optimization is not valid and will not be performed.

The compiler can apply the distributive, associative, and commutative laws in a complex fashion to discover common subexpressions.

Note that array syntax may prevent the compiler from saving results in registers. The following explicit loop code can be transformed using a scalar T that is stored in a register:

```
DIMENSION A(100), B(100), C(100), D(100), X(100), Y(100)
DO I = 1,100
    X(I) = A(I) * B(I) * C(I) * D(I)
    Y(I) = E(I) * B(I) * F(I) * D(I)
END DO
```

After common subexpression elimination, it becomes

```
DO I = 1,100
    T = B(I) * D(I)
    X(I) = A(I) * C(I) * T
    Y(I) = E(I) * F(I) * T
END DO
```

If the temporary scalar T is stored in a register rather than memory, this saves one multiply and two memory accesses per loop iteration.

If the user writes the code using array syntax rather than an explicit loop, the optimization looks similar to the scalar code above. But an array of length 100 must be allocated to hold the temporary T, which is now an array instead of a scalar or register. There is also a memory access for the store of the array T and the subsequent load; hence, 100 multiplies are saved but no memory accesses are.

EXERCISES

1. Try the following example with your compiler:

```
DO I = 1,100
DO J = 1,100
DO K = 1,100
    A(I,J,K) = B(I,J,K) + C(I,J,K)
ENDDO
ENDDO
ENDDO
```

And then

```
DO K = 1,100
DO J = 1,100
DO I = 1,100
    A(I,J,K) = B(I,J,K) + C(I,J,K)
```

```
ENDDO
ENDDO
ENDDO
```

Does your compiler perform this optimization automatically?

2. Given the Fortran 90 Array syntax example in this chapter, if the decomposition of parallel chunks were on the first and second dimension of the grid, at what sizes would the data being accessed in the code fit in Level 2 cache. Assume that the Level 2 cache is 512 KB and that the operands are 8-byte reals.

3. Why might the use of derived types degrade program performance?

4. Derived types do not always introduce inefficiencies; how would you rewrite the derived type in the example given in this chapter to allow contiguous accessing of the arrays?

5. Would $X = Y^{**}.5$ run as fast as $X = SQRT(Y)$? What would be a better way of writing $X = Y^{**}Z$? Try these on your compiler/system.

6. Why might the use of array syntax degrade program performance?

7. What constructs restrict the ability of the compiler to optimize the alignment of arrays in memory? What constructs gives the compiler the most flexibility in aligning arrays in memory?

8. What speedup factor is available from SSE instructions? How about GPGPUs?

9. Why is dependency analysis required to vectorize a loop?

10. What are the two methods for vectorizing conditional statements?

11. Why might a DO loop get better performance than the equivalent array syntax?

12. When using array sections as arguments to a subroutine, what may degrade performance?

13. What is strength reduction? How can it be used to speed up array index calculations?

14. What is hardware prefetching? What might the hardware examine to figure out what lines to prefetch.

15. What is software prefetching? How might it help the performance of a DO loop? What characteristics of an array might cause the compiler to do a prefetch on it?

16. If a user fine tunes the lengths of a DO loop to fit well in cache, would prefetching still be needed? Can prefetching hurt the performance of such a loop? How?

CHAPTER 4

Parallel Programming Paradigms

T HERE ARE NUMEROUS PARALLEL PROGRAMMING PARADIGMS being used today for both shared and distributed memory parallel programming. Each of these paradigms has their strengths and weaknesses. In this section, we will cover only those programming paradigms being used extensively in the HPC community.

4.1 HOW CORES COMMUNICATE WITH EACH OTHER

The microchip vendors supply multisocket nodes. For example, the Magny-Cours [18] chip from Advanced Mirco Devices (AMD) contains 2 four or six core sockets on a die. On a two die system, there can be as many as 16 or 24 cores sharing memory. When the user chooses to employ Message Passing Interface (MPI) across all of the cores, the system is treated as 24 separate processors (on the 2 die, 2 six core socket) with separate memories. Alternatively, if the user chooses to run OpenMP™ across a subset of the cores, that subset will share memory. A completely general operating system (OS), such as Linux®, allows any number of cores to be grouped as a single process with *n* parallel threads.

4.1.1 Using MPI across All the Cores

Running MPI across all the cores of a large massively parallel processor (MPP) system is the most prevalent programming strategy in HPC applications. Later in this section, we discuss some of the disadvantages of using this approach; however, the largest advantage is that a user can take their standard MPI job and run it across all of the cores in the MPP without changing anything. Surprisingly, most applications perform well with this simple approach. As we will see, when a user chooses to incorporate OpenMP and/or Pthreads; a.k.a. the hybrid approach, they are immediately faced with yet another programming paradigm—shared memory parallelization. Users are advised to see how much they can get out of using just MPI before making the leap to the hybrid approach.

Some vendors have developed optimized MPI libraries that are aware when shared memory operations can be used for messaging on the node. In this case, the MPI libraries would be able to optimize point-to-point communication when both MPI tasks are on the node. Ideally, the MPI library would pass pointers to the appropriate message buffers, so that the message never has to touch the interconnect. Such a library would be able to significantly reduce the latency and increase the bandwidth of messages whose origin and destination is on the same node. To see if a particular system has an optimized MPI runtime, the user can run a simple MPI point-to-point communication test when the two tasks are on the same node and then compare this to the case when the two tasks are on separate nodes. One should see a significant difference for small messages and less of a difference for very large messages.

Some MPI libraries have optimized collective communication such as global reductions and MPI_ALLTOALLs for multicore nodes. In collectives, the library usually uses a combining tree algorithm which can take advantage of the number of MPI tasks on the node. If a node has eight MPI tasks, then three levels of the reduction tree can be performed with memory copy operations very quickly and then the remaining levels of the tree will use the interconnect. These efficient MPI libraries significantly extend the value of the all-MPI approach.

4.1.2 Decomposition

The scalability of MPI is extremely dependent on the way the problem is decomposed across the processors. For example, if a 3D finite difference application is decomposed on 2D planes, then the maximum number of MPI tasks that can be effectively utilized is limited by the number of planes. This "one-dimension" decomposition is very limited. If, on the other hand, the application is decomposed on pencils; that is, one dimensional (1D) lines in the grid, then the number of MPI tasks that can be effectively utilized is much larger. The most versatile decomposition is when 3D subblocks that completely cover the original grid are used to

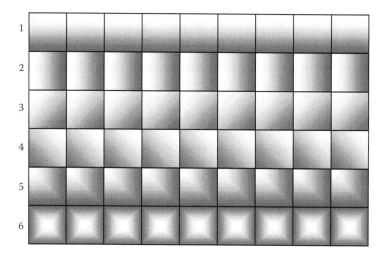

FIGURE 4.1   One dimensional only pencils decomposition.

decompose the problem. When selecting a decomposition for a large 2- or 3D scientific application, one wants to minimize the amount of communication between different MPI tasks. The surface area of a decomposed portion grid is directly proportional to the amount of data that must be transferred from the MPI task to the surrounding neighbors.

For a simple 2D example, consider the following very simple 2D grid. The first decomposition is to give each processor a line or a pencil of the grid. Each grid block must transfer a message to its right, left, up, and down neighbor (Figure 4.1).

If the grid is decomposed into horizontal pencils across six processors, then the interior processors have to send nine messages north and nine messages south—18 messages in total. If, on the other hand, a 2D decomposition is used, an interior MPI task would send three messages right, left, north, and south for a total of 12 messages in total. When the surface area of the decomposed chunk is made smaller, the number of messages to be sent can be reduced. When a 3D grid is used, a cubical decomposition will minimize the surface area. We will see later how such decompositions can take advantage of the alignment of the MPI tasks on the multicore node (Figure 4.2).

4.1.3 Scaling an Application

When looking at MPI programs, there are two common ways to scale an application to larger and larger processor counts. The most straightforward

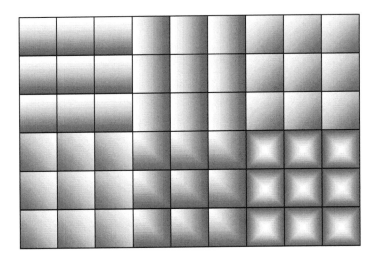

FIGURE 4.2   A 2D decomposition.

is keeping the problem size the same and as the number of processors increases, the amount of work that is performed by each processor decreases. This approach is referred to as "strong scaling." Even with the best decomposition, strong scaling will soon stop scaling when the amount of work performed by each processor is so small that the communication overwhelms the computation time. The alternative to strong scaling is weak scaling. "Weak scaling" refers to the approach of increasing the problem size as the number of processors increase. This approach tries to keep the amount of work per processor constant. The better the decomposition, the higher the weak scaling can scale. While weak scaling is still limited by the number of domains, the computation within a domain will remain constant as the number of processors increase; so communication should not overwhelm the computation. A word of caution when considering weak scaling: if increasing the complexity of the problem does not make good scientific sense, then it should not be used. Oftentimes, weak scaling can be used to illustrate how an application can scale to 100s of thousands of processors without doing any useful work. There are situations where weak scaling is ideal. For example, the S3D combustion code [8] requires an extremely fine mesh to perform the computations required to analyze their science. In this case, decreasing the grid dimensions and increasing the number of grids as the number of processors increase deliver excellent scaling and improve scientific discovery.

4.2 MESSAGE PASSING INTERFACE

The most widely used parallel programming paradigm is MPI. It is being used to successfully parallelize applications across hundreds of thousands of processors. A large majority of the MPI applications use what is known as single program, multiple data (SPMD) programming paradigm. The same program is run on every processor and each processor has its own data to compute. In the SPMD program there is only PRIVATE data, that is, each processor has its own data and no other processor shares that data. There is nothing automatic about MPI; the programmer is completely responsible for communicating data, synchronizing between processors, and coordinating the processors on solving a parallel problem.

4.2.1 Message Passing Statistics

As with hardware counter data, the user should gather message passing statistics for their application. MPI supplies a software interface profiling tool that can be used to gather statistics on the time used by various MPI calls. In addition to the actual time used by the MPI call, the tools can obtain information about the sizes of the messages. Profiling tools can also be utilized to identify call sites within the application responsible for most of the communication time and the sizes of the messages issued from that call site. With this information, a user can quickly identify which MPI operation is the bottleneck, and the sizes of the messages for determining if the bottleneck is due to latency of the interconnect or the bandwidth. Additionally, from the call site, the user can identify if any overlap is taking place. As the number of MPI tasks grows, the communication will become an appreciable portion of the total runtime, overlapping communication with other communication and/or computation becomes extremely important. Consider the data in Table 4.1 from the execution of a communication intensive application.

This table from the CrayPat™ tool [11] supplied by Cray Inc. gives extremely important information about the MPI call that transfers most of the data in the computation. The MPI routine being used is mpi_isend and this call transfers the same number of messages as the Total program does; so it is transferring all the data in the program. Now we see that the "call chain" (Main calls step_mod_step which calls boundary_boundary_3d_dbl_ which calls boundary_boundary_2D_dbl_ which calls mpi_isend) results in 641,487,168 of the 2,241,686,276 messages. All the messages are greater than 256 bytes and less than 4096 bytes. Below the call chain of routines calling mpi_isend comes the load balancing

TABLE 4.1   MPI Sent Message Statistics by Caller

Sent Msg Total Bytes	Sent Msg Count	256B <= MsgSz <4KB Count	Function Caller PE [mmm] Thread=0="HIDE"
2241686276	1807767	1807767	Total
|2241686276	1807767	1807767	mpi_isend_
||2241683463	1807762	1807762	boundary_boundary_2d_dbl_
3||1069392806	862390	862390	boundary_boundary_3d_dbl_
4|||641487168	517314	517314	step_mod_step_
5|||			MAIN_
6|||			main
7|||||| 7370899200	4903290	4903290	pe.2203
7|||||| 540046080	456120	456120	pe.1349
7|||||| 0	-	-	pe.0

statement for that particular call. In this case, processor 2203 transfers the maximum number of messages, while processor 1349 transfers an average number of messages, and processor 0 does the minimum, which is zero. From this, we see that this application is extremely latency sensitive since the messages are so small and there is significant variability in the amount of messages sent across the processors, which will cause load imbalance. More discussion of these issues will be covered later.

4.2.2 Collectives

When a majority of the communication time within an application is used in a collective, MPI_ALLREDUCE, MPI_REDUCE, MPI_ALLTOALL, MPI_VGATHER, MPI_VSCATTER, and so on, there may be a scaling problem. Collectives take a larger and larger fraction of time as the number of MPI tasks is increased. This is because a collective is not a scalable operation. As the number of MPI tasks increase, the time to perform the collective operation increases inordinately, owing to increased data motion required by the collective. For example, the MPI_ALLTOALL is frequently used to perform 2D and 3D transposes. In a weak scaling study, as the number of processors double, the amount of data being transferred across the interconnect is more than double. For example, if we have a 3D grid, if each processor has a given 3D decomposition, when the application must perform a transpose, each processor must send most of its data to every other processor. When the processor count doubles, the amount of data sent increases by eight. If the application using collectives is in a strong scaling environment, the amount of data to be transferred across the interconnect remains the same; however, the number of messages increases. There are methods of improving collectives. However, one severe limitation of a collective is that it is blocking—that is, while the collective is progressing, nothing else can execute in parallel. Future versions of MPI will address nonblocking collectives.

One method of improving a collective is to replace it with point-to-point communication. With point-to-point calls, communication can be overlapped with other communication and computation. A very good example of this is a 3D fast Fourier transform (FFT) that requires transposes between the various 2D FFTs on the planes of the grid. When using collectives to perform the transposes, nothing can be overlapped with the transpose. If the 3D FFT were written to use point-to-point communication, the communication could be overlapped with the execution of the FFT. This is not an easy task; however, it could result in the communication completely hiding the computation of the FFT.

Another powerful technique is combining collectives whenever possible. Oftentimes, numerous global reduces are performed at each iteration of a time step. These reduces should be combined into a single call to MPI_REDUCE, increasing the number of elements to be reduced accordingly. This approach will require that the elements to be reduced are packed into a contiguous section of memory. The time to perform a MPI_REDUCE or MPI_ALLREDUCE for 100 variables is only slightly longer than the time to perform the operation for one variable.

4.2.3 Point-to-Point Communication

When a majority of the time is spent in a point-to-point communication, the developer must examine the cause of that increased time. If the size of the message that takes most of the time is very small, then the latency of the interconnect is the bottleneck, and perhaps all the small messages that go to a given MPI task can be combined into a single, larger message. If the size of the message that takes most of the time is large, then the interconnect bandwidth could be the bottleneck and perhaps the messages should be broken up into smaller messages and sent asynchronously overlapped with computation.

4.2.3.1 Combining Messages into Larger Messages

Given an interconnect with a latency l and bandwidth b, the time to transfer a message of size n is given by

$$\text{Time} = l + \frac{n}{b}$$

The graph shown in Figure 4.3 gives the time to transfer as a function of the size of the message given a latency of 5 μs and a bandwidth of 2GB/s.

Note that the time is dominated by the latency until the message size is well over 100,000 bytes. With such a high latency it is preferable to combine messages to be sent from processor k to processor j. By packing two or three messages, the time to perform the pack is traded off versus the elimination of several latencies. This is an excellent technique for improving scaling performance when the application typically sends a lot of small messages. Such applications are known as latency sensitive.

4.2.3.2 Preposting Receives

The most effective way of using MPI to communicate between processors is by using point-to-point messages overlapped with computation. For example, when performing a halo exchange between two processors, each processor should post the receives of the messages coming from the

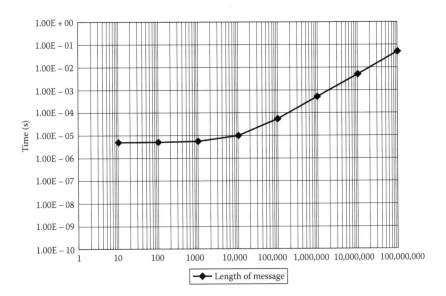

FIGURE 4.3 Time to transfer message.

neighboring processors prior to sending its data to the neighboring processors. Both the receives and sends can be nonblocking—that is, they do not wait until the operation is complete. Using this approach, the processors can overlap these asynchronous sends/receives with computation. Only when the processor needs the data from its neighbors does it need to use an MPI_WAIT to assure that the message is received.

When the receive is preposted, the message passing software has a buffer to place the message in when it arrives at the receiver. It simply places the message into the application buffer passed in the MPI call. When the message arrives before the receive is posted, the runtime routine must allocate a buffer to save the message until the receive is posted. This allocation of memory takes time and the additional transfer of the message, first to the temporary receive buffer and then to the application buffer, will take more time than the straightforward message receive into the application buffer when the receive is preposted. Since some processors may have the receive posted prior to receiving the message, this will introduce a load imbalance simply by some processors receiving the data faster than other processors. The impact of the inefficiency of not preposting receives is much more apparent at very large processor counts (>20,000 cores).

Most intelligent MPI libraries will prepost the receive when an MPI_SENDRECV is used; however, this is not a long enough time to prepost the message. Since the MPI tasks are operating independently, the likelihood of a processor receiving a message from another processor before it executes the MPI_SENDRECV is very high. Ideally, the receive is posted as soon as the buffer to contain the message is available.

In reviewing the point-to-point communication, the following questions should be asked:

1. What is the earliest time I can receive the data that I need from my neighbor?

   a. That is, the point where a receive (MPI_IRECV) should be posted

2. Do I need to allocate a separate buffer so that I can calculate on a previous message while I receive the next message for the next computation?

   a. Overlap communication with computation as much as possible

3. When can I send the data that the other processor needs?

4. When must I wait to make sure my receive is completed?

5. When must I wait to make sure my send is completed?

By answering these questions, the code developer can then develop an asynchronous communication scheme, where the receives are posted far enough ahead of when they are needed and computation can be performed while the receive is being performed. Also, consider overlapping communication with other communication as well as computation. Consider the following except of code from an application that simulates earthquakes:

```
call wu2ed_f(rank,comm,tru1,ix_d,fx_d,iz_d,fz_d,nx,nz,wu2ed, ed2wu)
call ed2wu_f(rank,comm,tru1,ix_d,fx_d,iz_d,fz_d,nx,nz,ed2wu, wu2ed)
call wu2ed_f(rank,comm,trw1,ix_d,fx_d,iz_d,fz_d,nx,nz,wu2ed, ed2wu)
call ed2wu_f(rank,comm,trw1,ix_d,fx_d,iz_d,fz_d,nx,nz,ed2wu, wu2ed)
call wd2eu_f(rank,comm,tru1,ix_d,fx_d,iz_d,fz_d,nx,nz,wd2eu, eu2wd)
call eu2wd_f(rank,comm,tru1,ix_d,fx_d,iz_d,fz_d,nx,nz,eu2wd, wd2eu)
call wd2eu_f(rank,comm,trw1,ix_d,fx_d,iz_d,fz_d,nx,nz,wd2eu, eu2wd)
call eu2wd_f(rank,comm,trw1,ix_d,fx_d,iz_d,fz_d,nx,nz,eu2wd, wd2eu)
```

Each of these calls perform communication with nearest neighbors. Within each of the calls there is a SEND/RECV pair.

```
if(dst /= MPI_PROC_NULL.and.fz_d==nz) then
    bto = b(1:fx_d,fz_d-1:fz_d)
    call MPI_SEND(bto,npts,MPI_REAL,dst,rank,comm,err)
end if
if(src /= MPI_PROC_NULL.and.iz_d==1) then
    call MPI_RECV(bfrom,npts,MPI_REAL,src,MPI_ANY_
        TAG,comm,status,err)
    b(1:fx_d,-1:0) = bfrom
end if
```

Note that all of these calls are blocking, there is no overlap of these calls even though the eight messages are independent. It is possible for all of these neighbor updates to be performed in parallel, by simply, initially making the eight calls to a routine that preposts the receives and then make eight additional calls that performs the sends and waits for the receives to be completed. When point-to-point communication is written to overlap communication with other communication and computation as illustrated in these examples, weak scaling problems that are dependent on point-to-point communication will never stop scaling. Strong scaling problems will benefit and scale until there is not enough work to hide the communication.

4.2.4 Environment Variables

Most systems will have environment variables that the user can set to control the operation of MPI during runtime. The most common is an MPI threshold variable that designates at what message size a message should be received prior to being requested by the receiver. For example, in a Halo exchange, a sending processor may send a message to its neighbor before the receiver has posted a receive. If this is the case, the receiver does not know where to put the data and it must allocate some memory to hold the message until the receiver posts the message. Alternatively, the receiver could respond to the sender to hold the message until the receive is posted so that it will have a location to put the message. Now the sender has to allocate some memory to hold the message or wait on the send until the receiver notifies it that it can send the message. Typically, small messages are always received and saved by the receiver and at a given threshold, the delay method is used. Along with the environment variable that specifies this threshold, the user typically needs to control the amount of memory allocated for the buffers required to hold the messages. Appendix B has a list of the MPI variable for the Cray XT™ system. Other systems probably have comparable environment variables that must be considered when fine tuning the scaling of an application.

4.2.5 Using Runtime Statistics to Aid MPI-Task Placement

When running on a multicore system, load imbalance in an application can be addressed with good MPI task placement. Consider an application where the amount of computation varies due to the science being computed. For example, early Ocean models included land in the grid and those processors that had land grids would do nothing. This decomposition was used to maintain a regular grid. Since the land areas are not uniform, a regular latitude–longitude grid would result in some processors having all ocean while other processors have mostly land. In these situations, it might be beneficial to group both the ocean and land grids on the same node in such a way that each node has the same number of land and ocean grids. Unfortunately, this grouping would probably result in nearest neighbors no longer being on adjacent nodes of the MPP. Such an irregular grouping of the work would have the impact of removing the load imbalance due to work load; however, it could increase the communication. On those systems that have sufficient interconnect bandwidth, the approach has shown to be valuable.

When the profiling tools supply information about load imbalance, this information can be used to gather MPI tasks on nodes to balance the amount of processing performed on a node. As in the previous example, this is a very powerful technique to minimize the degradation in performance due to a natural imbalance in the processing performed in an application. This imbalance may be computational or it can be communication. Using this manual, MPI task placement on the nodes can address communication imbalance as well. In irregular grids, some processors may have to perform more communication than others. As with the case of the imbalance computation in the previous example, the user should place the MPI processes that perform more communication with the processes that perform less communication. In this way the processes that perform more communication can obtain more of the network injection bandwidth.

4.3 USING OPENMP™

When the user takes an MPI application and starts to introduce OpenMP, they immediately reduce the number of MPI tasks by N (2, 4, or 8 on a dual quad-core system) and use N OpenMP threads per MPI task. Once OpenMP is employed, the application is using fewer MPI tasks and there will be an increase in wall-clock time due to fewer MPI tasks. This reduction in performance can be as much as a factor of N. That loss of a factor of N must be recovered by parallelizing with shared memory parallelism across the N cores. The speedup obtained by applying shared memory parallelism is dependent on the amount of computation parallelized with the OpenMP directives. The following chart illustrates the relationship between performance achieved versus the percent of computation that is parallelized. In the chart, we look at the percentage of computation that must be parallelized with OpenMP on a four-core socket to achieve a desired speedup. Notice that you need to parallelize 80–90% of the code to obtain a factor of three overall. This is not easy to do (Figure 4.4).

OpenMP is an evolving language extension that allows programmers to direct the compiler to parallelize sections of code across processors that share memory. This shared memory can be either a flat memory space, such as on most single-socket multicore nodes or a nonuniform shared memory where some portion of the memory is closer to one processor than others. When using a multicore node, it is extremely important to group the OpenMP threads on the right socket. All shared memory systems have the ability to bind OpenMP threads to specific cores.

In Chapter 8, several OpenMP applications are examined and their overall performance gain will be shown to be extremely dependent on the amount of code parallelized.

Introducing mixed OpenMP and MPI can allow many applications to scale to higher core counts. When investigating mixed OpenMP and MPI, it does not make sense to compare the mixed version of an application to the all-MPI version in a situation where the MPI version is scaling. Introducing OpenMP into an MPI application should be performed at the point where MPI has quit scaling due to problems with the decomposition, and so on.

4.3.1 Overhead of Using OpenMP™

There are several types of overhead associated with OpenMP, which is a master–slave paradigm. A single core will execute the serial part of the application and then when a parallel section is encountered there will be overhead associated with requesting the other cores to share the work in the parallel region. Then, at the end of the parallel region, only the master thread will continue working on the serial code. For this reason, it is extremely important that the parallel regions take significant time. Combining parallel regions together to increase the granularity of the parallel work can be very beneficial. In Chapter 8, we also examine the granularity of the loops being parallelized and demonstrate that the larger the granularity, the better the speedup for parallelizing the DO loops with OpenMP. OpenMP is a set of comment line directives that are used

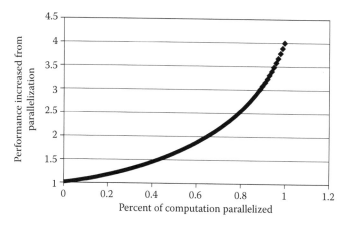

FIGURE 4.4   Amdahl's law applied to parallelism on the socket.

to designate what to parallelize, how to scope the variables used within the parallel region, and finally, how to distribute the parallel work across the processors.

4.3.2 Variable Scoping

Perhaps the most difficult process associated with the use of OpenMP is the process of scoping the variables that are used in a parallel structure. When parallel processors (threads) execute within a parallel region, they must have some amount of data that is private to a particular thread. Then there are data that are shared among the threads. When a compiler automatically parallelizes a DO loop, one of the functions it must perform is variable scoping and then allocate the appropriate private data for each thread. The rules of scoping are quite simple; however, in practice, they are difficult to implement.

1. Any scalar variable or array constant that is set and then used each pass through the DO loop should be private to each thread. A good example is the DO loop index.

2. Any scalar variable or array constant that is used and then set in a single loop pass should be shared—this will cause a loop carried dependency.

3. Any scalar variable or array constant that is used and then set in a single line is a shared variable, and may be a reduction function scalar.

4. Any variable that is only read in a DO loop is shared.

5. Any array referenced by the parallel loop index is shared.

These rules are quite simple; however, complications arise when the OpenMP DO loop encompasses subroutine calls and the scoping of a variable down the call chain does not adhere to language storage characteristics. For example, say a variable—work_temp—is in a module that is used within a routine that contains a DO loop as follows:

```
DO I=1,N
    WORK_TEMP = SQRT(A(I)**2+B(I)**2)
    ARRAY(I,J) = WORK_TEMP+ARRAY(I,J)
ENDDO
```

Given the scoping rules, WORK_TEMP should be private; however, since WORK_TEMP is in a MODULE, there is only one memory location associated with it. Some cases are easy to handle and the most compilers will allow the programmer to scope WORK_TEMP private. In this case, the compiler ends up dynamically allocating a WORK_TEMP for each thread and then saving the WORK_TEMP value set by the last pass through the loop into the MODULE storage location. This is known as last value saving. This is a simple case that can be handled automatically, other cases are not so easy. When the module variable is used in a subroutine and it is set at each iteration of the OpenMP loop and then used, the user must handle this situation by restructuring the routine of the module.

The most benefit from parallelizing with OpenMP is obtained when the parallel DO loop encompasses a large amount of computation. When moving up the call chain to find a parallel DO loop, scoping issues are significantly complicated. Consider the following situation:

```
DO I=1,N
    CALL CRUNCH(A,B,C,D)
ENDDO
```

The only variables that can be scoped on this loop are I as private and N as shared. CRUNCH needs to be examined to determine the scoping of A, B, C, and D and additional global rules come into effect.

1. Scoping of variables down the call chain must assume the scoping rules of FORTRAN

    a. Shared variables must be GLOBAL variables

        i. In MODULES

        ii. In COMMON blocks

        iii. Arguments passed by be private depending on the calling routine

    b. Private variable must be LOCAL variables

        i. ALLOCATED variables

        ii. AUTOMATIC variables

        iii. Arguments passed may be shared depending on the calling routine

2. TASK_COMMON is supported by some compilers to handle COMMON block variables that must be made private to be able to parallelize the DO loop.

3. OpenMP has a THREADPRIVATE directive to make global file scope variables (C/C++) or COMMON blocks and MODULES (Fortran) local and persistent to a thread.

So while the rules are simple, the implementation can be quite difficult, especially when OpenMP is used at a high level, where it should be used.

4.3.3 Work Sharing

The next task the user should consider is how to distribute the parallel work across the threads on the node. The default on DO loop parallelization is typically to distribute the loop iterations to each thread in a STATIC mode. By STATIC we mean that the loop iterations are distributed evenly to the threads at the top of the loop. The other options are to distribute iterations of the loop in a dynamic mode and/or specify the size or number of iterations that should be allocated to a thread.

When the compiler performs default work sharing, it assumes that all the iterations of the DO loop take equal time. When that is not the case, the user should consider a more dynamic work sharing construct. For example, using dynamic scheduling, when a thread finishes its chunk of work it is assigned the next available chunk of work. In this situation, some threads may do more work than others. This method effectively distributes the work across the threads by dynamically allocating the loop iterations to the next available thread. The allocation of DO loop iterations brings up an interesting trade-off. Since the assignment of work to a thread takes some amount of overhead, the amount of work assigned should be relatively large chunks. If the iterations of a DO loop vary greatly in time to execute, then the work should be allocated with one iteration at a time. The smaller the chunk of work or granularity, the larger the proportion of overhead. Therefore, there is a delicate trade-off between load imbalance, which takes away performance and overhead of issuing a work schedule for the threads, which also takes away performance. The user must understand the application and try to devise large enough granules of work without causing significant load imbalance.

Still another question arises as to whether the work is assigned STATICally or DYNAMICally. Work allocation is assigned to the next available thread either STATICally at DO loop initiation and does not change during the execution of the DO loop or DYNAMICally at runtime. While DYNAMIC has more overhead it does handle nonload-balanced loops very well.

OpenMP tends to have large overhead and the granularity of the parallel DO loop should be very large. There are ways of increasing the granularity of many small parallel DO loops by introducing higher level DO loops that strip mine across the inner loop. In this way, the many small parallel DO loops can be combined together into a larger parallel loop. This is particularly useful when trying to parallelize many FORTRAN 90 array assignments. Consider the following coding excerpt from the S3D combustion code.

```
 !--update error
!$OMP Parallel DO
  q_err(:,:,:,:,1) = q_err(:,:,:,:,1) + rk_err(jstage)
                    * tstep * q(:,:,:,:,2)

 !--update solution vector
!$OMP Parallel DO
  q(:,:,:,:,1) = q(:,:,:,:,3) + rk_alpha(jstage)
                * tstep * q(:,:,:,:,2)

 !--update carry-over vector
!$OMP Parallel DO
  q(:,:,:,:,3) = q(:,:,:,:,1) + rk_beta(jstage)
                * tstep * q(:,:,:,:,2)
```

The OpenMP directives are placed on each individual statement. However, introducing a new loop on the third dimension and using OpenMP on that loop is more efficient due to less overhead.

```
!$OMP Parallel DO
DO K=1,nz
   !--update error
   q_err(:,:,k,:,1) = q_err(:,:,k,:,1) + rk_err(jstage)
                      * tstep * q(:,:k:,:,2)
   !--update solution vector
   q(:,:,k,:,1) = q(:,:,k,:,3) + rk_alpha(jstage)
                  * tstep * q(:,:,k,:,2)
   !--update carry-over vector
   q(:,:,k,:,3) = q(:,:k:,:,1) + rk_beta(jstage)
                  * tstep * q(:,:,k,:,2)
END DO
```

This will give us larger granularity in that there is more work within the parallel DO loop.

4.3.4 False Sharing in OpenMP™

When an OpenMP loop index accesses the leftmost contiguous dimension in an array (left in Fortran, right in C), it is likely that different threads will simultaneously access the same cache line, severaly degrading performance. Consider the following DO Loop:

```
!$OMP PARALLEL DO PRIVATE(I,II,IBD,ICD)
      DO I = 2,NX-1
        II = I + IADD
        IBD = II-IBDD
        ICD = II + IBDD
        DUDX(I) =
>        DXI * ABD *((U(IBD,J,K) - U(ICD,J,K)) +
>            8.0D0 *(U( II,J,K)   - U(IBD,J,K)))* R6I
        DVDX(I) =
>        DXI * ABD *  ((V(IBD,J,K)- V(ICD,J,K)) +
>            8.0D0 *(V( II,J,K)   - V(IBD,J,K))) * R6I
        DWDX(I) =
>        DXI * ABD *((W(IBD,J,K) - W(ICD,J,K)) +
>            8.0D0 *(W( II,J,K)   - W(IBD,J,K))) * R6I
        DTDX(I) =
>        DXI * ABD *((T(IBD,J,K) - T(ICD,J,K)) +
>            8.0D0 *(T( II,J,K)   - T(IBD,J,K)))* R6I
      END DO
```

Say we have four threads working on this OMP loop and NX is 200. To divide the loop into four chunks we calculate $(199 - 2 + 1)/4 = 49.5$. Since we cannot give out an noninteger number of iterations, the first two threads will get 50 and the last two will get 49. When thread 0 is storing into DUDX(51), thread 1 may be storing into DUDX(52). There will be a problem if both of these elements of the DUDX array are contained on the same cache line. At these intersections between the iterations between two threads, we could have multiple threads fighting over the ownership of the cache line that contains the elements of the arrays that each thread wants to store into.

Evidently, one should never use OpenMP on the contiguous index of an array and/or if they do they should take care to distribute iterations to each of the threads according to cache lines rather than the range of the

DO loop. This is only a concern when storing into an array, and in this case, it may be very difficult to divide the iterations; hence, all four of the stores are avoiding overlapping requests from the same cache line. False sharing also occurs when several threads are writing into the same cache line of a shared array. Another application from the SPEC_OMP benchmark suite has this OpenMP parallel loop:

```
C
      DO 25 I=1,NUMTHREADS
        WWIND1(I)=0.0
        WSQ1(I)=0.0
  25    CONTINUE

!$OMP PARALLEL
!$OMP+PRIVATE(I,K,DV,TOPOW,HELPA1,HELP1,AN1,BN1,CN1,MY_
      CPU_ID) MY_CPU_ID = OMP_GET_THREAD_NUM() + 1
!$OMP DO
      DO 30 J=1,NY
        DO 40 I=1,NX
          HELP1(1)=0.0D0
          HELP1(NZ)=0.0D0
          DO 10 K=2,NZTOP
            IF(NY.EQ.1) THEN
              DV=0.0D0
                  ELSE
              DV=DVDY(I,J,K)
            ENDIF
            HELP1(K)=FILZ(K)*(DUDX(I,J,K)+DV)
  10      CONTINUE
C
C    SOLVE IMPLICITLY FOR THE W FOR EACH VERTICAL LAYER
C
        CALL DWDZ(NZ,ZET,HVAR,HELP1,HELPA1,AN1,BN1,CN1,ITY)
        DO 20 K=2,NZTOP
          TOPOW=UX(I,J,K)*EX(I,J) + VY(I,J,K)*EY(I,J)
          WZ(I,J,K)=HELP1(K) + TOPOW
          WWIND1(MY_CPU_ID)=WWIND1(MY_CPU_ID)+WZ(I,J,K)
          WSQ1(MY_CPU_ID)=WSQ1(MY_CPU_ID)+WZ(I,J,K)**2
  20      CONTINUE
  40    CONTINUE
  30  CONTINUE
!$OMP END DO
```

```
!$OMP END PARALLEL

      DO 35 I=1,NUMTHREADS
         WWIND=WWIND+WWIND1(I)
         WSQ=WSQ+WSQ1(I)
 35   CONTINUE
```

Notice that the WWIND1 and WSQ1 arrays are written into by all the threads. While they are writing into different elements within the array, they are all writing into the same cache line. When a thread (a core) updates an array, it must have the cache line that contains the array in its Level 1 cache. If a thread (a core) has the cache line and another thread needs the cache line, the thread that has the line must flush it out to memory so that the other thread can fetch it up into its Level 1 cache. This OpenMP loop runs very poorly. An improvement is mentioned in Chapter 8.

4.3.5 Some Advantages of Hybrid Programming: MPI with OpenMP™

There are several reasons why an application programmer should move to the Hybrid programming paradigm that combines MPI with OpenMP. Perhaps the most evident is that as multicore sockets and nodes grow larger and larger, the impact on the interconnect increases as applications continue to run MPI across all the cores of a system. A second less obvious reason is that future accelerators will require shared memory parallelism combined with low-level vectorization. Moving to those new architectures will require some type of hybrid programming.

4.3.5.1 Scaling of Collectives

The biggest disadvantage of the all MPI approach is that as the number of MPI tasks increase, the communication required to perform the collective increases inordinately. If, on the other hand, OpenMP is added to extend the scaling of an MPI application, the collective communication does not increase. Numerous applications that employ 3D FFTs which perform transposes across the grid have successfully introduced OpenMP to scale to higher and higher core counts.

4.3.5.2 Scaling Memory Bandwidth Limited MPI Applications

Many MPI applications are memory bandwidth limited and the user may choose to run on a subset of the cores on a multicore socket. Higher overall performance per core will always increase as the number of cores used on the socket is reduced. This is due to reducing the pressure on the memory bandwidth. When this approach is taken there are idle cores on the socket. If this is the case, adding OpenMP to utilize these idle cores should improve the scaling. If the OpenMP is done poorly and impacts the memory bandwidth adversely, then this approach would not be valid.

4.4 POSIX® THREADS

A method of implementing parallelism on shared memory multiprocessors is the use threads. A thread is an independent unit of execution. The execution of threads on available processors is controlled by the operating system scheduler. When multiple cores are available, the scheduler will put different threads on different cores so as to keep as many cores busy as possible. If there are more threads than cores (and all the threads are ready to run), all cores will be used simultaneously. This is one way of maximizing the result rate of a multicore CPU.

An obvious method of using all the cores on a CPU is to run several programs simultaneously. Posix® threads are a means for dividing a single program into multiple pieces and running the pieces simultaneously. Threads have been used on single processors to allow rapid response by interactive users. A time-consuming operation can be performed by one thread while a different thread interacts with the user. The scheduler runs the CPU-intensive thread while waiting for user interaction with the user interface. When the user interacts with the mouse or keyboard, the scheduler quickly suspends the compute-intensive task and resumes the interface task. This is largely invisible to the user. This is particularly useful for programs using a graphic user interface. The user gets nearly instant response from one thread while computations continue in the background by the other thread.

Threads generally have less overhead than other multiprocessing methods such as the Unix® "fork" mechanism. Vendor-specific thread libraries have been superceded by a standard named Posix Threads available on most operating systems. The name "Pthreads" is a common name for this implementation.

Parallel execution on shared memory multiprocessors is enabled by the Pthread library. If the number of threads is equal to the number of processors, the scheduler will execute each thread on a separate processor.

The first step in the parallelization of a program with Pthreads is to determine how the program computations will be divided among the threads. There are three general designs for partitioning calculations among threads. They are

- Functional decomposition

- Data decomposition

- Pipelining

Functional decomposition assigns different functions to different threads. Using the analogy of building a house, one thread makes the sidewalks, another installs the plumbing, and a third puts on the roof. This is appropriate when the different functions are largely independent.

Data decomposition assigns the same function to different threads, but each thread uses a different section of data. Using the analogy of building a house, one thread installs the wiring in the living room, while a second thread installs the wiring in a bedroom, and a third installs the wiring in the kitchen. When all the wiring is finished, each thread installs the drywall in its room. Then each thread paints the walls in its room. This continues until each thread finishes its room. The most common construct for data decomposition in a parallel program is a loop. Each iteration of a loop can be assigned to a different thread.

Pipelining is similar to functional decomposition. It is used when the functions are not independent and must be performed sequentially on each set of data. Again, using the house building analogy, think of three workers. One installs wiring, second installs drywall, and third paints the walls. In each room, the wiring must be installed before the drywall is installed. The drywall must be installed before the walls can be painted. The pipeline method sends the electrician to install the wiring in the first room while the drywaller and painter are idle. When the electrician finishes the first room, he proceeds to the second room, while the drywaller starts on the first room. The painter is still idle. When the electrician finishes the second room, he moves to the third room. The drywaller finishes the first room and moves to the second. The painter then starts on the first room. The three workers, or threads, proceed, each doing a room at a time, until the electrician finishes the last room. The electrician is idle while the drywaller works on the last room and the painter works on the next to the last room. When the drywaller finishes the last room, he and the electrician are idle while the painter paints the last room.

These techniques can be combined. Data decomposition can be used to assign a different set of workers to each of multiple houses to be built. The set of workers assigned to a particular house can use functional decomposition or pipelining, or a combination, to work on a single house in parallel.

The goal in choosing a parallelization strategy is to achieve maximum speedup from the available cpus or cores, with a minimum amount of programming work writing, debugging, and maintaining code. The strategy may be different if a parallel program is written from scratch. When modifying an existing serial code it is usually desirable to use existing code, inserting a parallel framework to execute serial pieces with threads.

The strategy for parallelizing a program depends on multiple factors. Some of these factors are

- Granularity

- Interthread dependencies

- Size of data sets

- Load balancing

- Memory contention

A straightforward parallel implementation simply starts one thread to perform each unit of parallel work, then waits for all of them to finish. This is repeated for each section of code that contains parallel units of work. The main program starts the threads, waits, and then continues to the next section of the code.

A potential problem with this approach is the overhead creating, starting, and stopping threads. This is typically a few milliseconds. If the work performed in a parallel section also takes a few milliseconds, most of the time will be spent in overhead starting and stopping threads rather than useful work. A second method is to use a pool of threads. A number of threads are created, and are then left waiting for something to be done. When the main program reaches a parallel unit, it initializes a data structure for each thread containing pointers to code and data sufficient for each thread to perform its computations. The main program then sets a condition for each thread telling it to GO. Each thread waits for its condition. When the condition is set, the thread performs the work described by its data structure. The main program waits for each thread to set a condition indicating that it is finished. When finished, each thread sets the condition for the main program and then waits for the next condition to be set when there is more work to do. If a pool of running threads is used, there is less overhead for each parallel unit of work, but there is still overhead.

The amount of time spent by a single cpu or core executing a thread from start to finish is termed the granularity. Low or fine granularity means that a short time, a few milliseconds, elapses between the start and stop of a thread execution. High or coarse granularity means a large amount of time elapses. The higher the granularity, the more time is spent performing useful calculations instead of overhead code starting and stopping threads, not required by the serial code.

When multiple threads execute simultaneously, they often have a different amount of work to perform. If five painters are tasked to each paint one of five rooms, the one painting the smallest room finishes first. He may then sit idle while the other four finish their room. Then the one painting the next smallest room finishes and is idle. This proceeds until the last one painting the largest room finishes. If the largest room is much larger than the rest, four of the painters will be idle most of the time. Attempting to give all the painters, or all the threads, an equal amount of work to perform to avoid idle time is called load balancing.

If the number of parallel tasks to perform is equal to the number of cores, the only means to achieve load balance is to design the parallel algorithm so that the size of the parallel tasks is as equal as possible. (Size means amount of work to perform, not the number of lines of code.) If there are more parallel tasks to perform than there are threads, the tasks can be divided among the threads to maximize load balance. If there are two workers to paint two small rooms and one large room, one painter does the large room while the other does both small rooms. Generally, if there are many more parallel units than threads, it is easier to achieve good load balance.

Often, the amount of work required for each parallel task is difficult to predict. In this case, dynamic load balancing can be used. Each thread is given a task. Whenever a thread finishes a task, it is assigned a new task to execute. This continues until all tasks are finished.

Memory access patterns can drastically affect the performance of multithreaded code. When two cores are executing simultaneously, each accesses the same global memory. The hardware and software assure that it works. If one core sets a memory location and the other core later reads it, the second core reads the value set by the first core. The existence of cache on both cores makes this memory synchronization complicated. As stated elsewhere, memory is divided into cache lines of consecutive words of memory. When different cores set and access the same cacheline nearly simultaneously, the hardware operates as if they are accessing the same word in memory. Cache must be written to memory and invalidated on one or both cores. This is very time consuming and will destroy performance if it occurs many times. A common cause for this is using data decomposition on the wrong index of an array. The following illustrates this:

```
REAL A(2,1000,1000)
DO K=1,2
  DO J=1,1000
    DO I=1,1000
      A(K,J,I) = . . .
    ENDDO
  ENDDO
ENDDO
```

If each iteration of the K loop is performed by a different thread, and a different core, performance will be horrible. A(1,K,J) and A(2,K,J) are adjacent in memory and almost certainly in the same cacheline. If this is rewritten as below, performance is better.

```
REAL A(1000,1000,2)
DO K=1,2
  DO J=1,1000
    DO I=1,1000
      A(J,I,K) = . . .
    ENDDO
  ENDDO
ENDDO
```

Sometimes each thread can execute from start to finish independently without interference from other threads. But some algorithms require that the threads synchronize execution to perform calculations in an order that generates correct answers. There are two forms of synchronization commonly encountered.

The first form of synchronization is the prevention of interference between threads. A common construct requiring this synchronization is the increment of a counter. Consider an algorithm that, in addition to time-consuming computations, also counts the occurrence of one or more conditions. A thread increments one or more counters with a simple line of code:

$$N = N + 1$$

If two threads execute this code simultaneously there is a chance that both threads will fetch N, both will increment it, and then both will store it, with the consequence that the value of N stored has only been incremented once, not twice, as intended. This error is particularly troublesome because it is dependent on the timing of the threads. As long as the threads do not execute in this exact sequence, the result will be correct. Errors will be intermittent and not reproducible. A construct called mutex (mutual exclusion) locks is implemented in the Pthread library to prevent this interference. A call to a LOCK routine is placed before $N = N + 1$. A call to UNLOCK is placed after it. The LOCK routine checks if any other thread has executed the LOCK, but not the UNLOCK. If not, it "locks" the section, performs the assignment, and unlocks the section. If the check reveals that another thread does have the section "locked," the thread waits until the other thread "unlocks" the section. Interference is thus avoided. LOCKs involve overhead. The code to perform the locking and unlocking is overhead. Time spent by one thread waiting for another thread to unlock a section of code is also overhead. Locks should only be used when necessary. Algorithms should be designed to minimize or combine locks. The section of code between the lock and unlock should be as small as possible.

A second form of synchronization is required when one thread must finish a part of its work before another thread can start. In the functional decomposition method in the house worker analogy, the drywaller cannot start work in a room until the electrician is done. The Pthread library includes routines to create, initialize, set, waitfor, and destroy conditions. The SET and WAITFOR routines perform the actual synchronization. A condition, intially false, is associated with each room. The drywaller waits for a rooms condition to be set to true before working in that room. The electrician sets the condition to true when he finishes the wiring for the room.

Testing and debugging a Pthread parallel implementation has several problems. First, the code must be correct without considering parallelism. It is often best to start with a working serial program, or to write a serial version first. When possible, the parallel program should be tested using a single thread. This may not be possible in some forms of functional decomposition or pipelining which require a different thread for each function. Debugging can often be performed in a machine with a single core, or fewer cores than will ultimately be used. In general, there can be more threads than cores. The parallel program will execute more slowly, but will finish. Debuggers, such as TotalView, support debugging parallel threads. Intermittent, nonrepeatable, time-dependent bugs, usually caused by the lack of or misuse of locks and conditions, are very difficult to find. A program may produce incorrect results one time in 10. The incorrect results may be different each time. It is often impossible to produce the bug on demand. Such debugging is best avoided by being very careful designing and writing parallel code to know what synchronization is required and performs the synchronization correctly.

4.5 PARTITIONED GLOBAL ADDRESS SPACE LANGUAGES (PGAS)

MPI has been the principal interface for programming applications for MPP systems. As the size of the MPP system grows and as applications want to scale to larger processor counts, we see limitations in MPI. The biggest limitation is the need for a handshake between the sender and the receiver of each and every message. In Chapter 2, the handling of MPI messages by the Network Interface Computer (NIC) was discussed. As the number of cores on the node increases, the number of messages coming off the node increases and soon the message matching process starts to limit the performance of the applications. Some interconnects have the ability to perform a get from another processor's memory or a put into another processor's memory without the other processor performing a receive or a send. Such a messaging capability eliminates the need to match messages between the sender and the receiver. Synchronization around the remote gets or puts is the responsibility of the application developer. With the capability to perform direct gets or puts into a remote processor's memory, a new programming paradigm has evolved called "PGAS" languages. The major PGAS languages at this time are co-array Fortran (CAF) [9] and Unified Parallel C (UPC) [10]. Both of these languages give the programmer extensions to FORTRAN and C, respectively that allow direct access to the memory of another processor. A significant advantage of these languages is the ability to incrementally modify an existing

FORTRAN/MPI and/or C/MPI applications by replacing MPI constructs with what might be more effective communication. There is no need to replace all the MPI calls, just those operations where the ability to use asynchronous gets and puts would improve the performance.

In order to run a CAF and/or UPC program efficiently on a large multicore architecture, both the compiler and the interconnect must be PGAS friendly.

4.5.1 PGAS for Adaptive Mesh Refinement

When remote get and put operations are available, different strategies can be investigated for solving difficult message passing situations. Consider an application that needs to send messages to a nonuniform collection of processors. Rather than a nearest-neighbor scenario, consider the case where the location and number of processors that must be communicated with is totally random and changes as the computation grows. Such an application was very difficult to develop in MPI efficiently. Such might be the case in many adaptive mesh refinement codes and particle in cell codes. There was a process that determined what processors were now neighbors; then the messages would be packed in a buffer to gain the benefit of transferring larger messages, then send the data, receive messages from its neighbor, and finally unpack the incoming messages into the appropriate arrays.

With a PGAS language such as UPC and/or co-arrays, the logic to figure out what processors must be communicate to still needs to be done; however, when the data are available to send to the distant processor, a small message can asynchronously be shipped off to the destination. PGAS is excellent for getting a large number of small messages in flight at the same time. This overlap of sending messages is one of the benefits of PGAS.

Of course the application developer must then be careful to check for the incoming messages, making sure they have arrived before the data are used. Additionally, the processor must make sure that the outgoing message was received prior to reusing the buffer space for the next set of messages. The synchronization of PGAS languages is much more complex than when using MPI.

4.5.2 PGAS for Overlapping Computation and Communication

Today many applications must perform very message-intensive collective communication and then perform a major computation. For example, computing a 3D FFT requires several MPI_ALLTOALLs and FFT computations on the processors. Today MPI_ALLTOALL is blocking; that is, all activities must stop until the communication is complete. The next MPI standard may introduce nonblocking MPI_ALLTOALL, where the processor can perform some computation while the communication is taking place. A better alternative is to incorporate PGAS within the FFT computation to start sending data to the other processors as the FFT is being computed. It is efficient to send numerous small messages overlapping the communication and overlapping the communication with the computation of the FFT.

These two examples of where PGAS can outperform MPI are just a sample of the possibilities. Whenever the communication is irregular, consisting of small messages, able to be overlapped with computation, PGAS should be investigated. Unfortunately, PGAS will not be performance portable for some time. A user may write a perfect PGAS program and the target system may run it much slower than the equivalent MPI program. Two additional elements must be in place to assure the efficient execution of a PGAS program. When the compiler understands the implications of the PGAS syntax, it should be able to perform pre-fetching of the remote memory reference, just as it does for a local memory reference. Even to the point of performing prefetching on the variables. Next, the interconnect must be a true remote direct memory access system. In some cases, the data might even be placed in the cache of the target put, or obtained from the cache of a target get.

4.5.3 Using CAF to Perform Collective Operations

When an MPI_ALLTOALL is called in an application program, all computation and other communication must come to a halt and wait until the operation is complete. While there are discussions about adding nonblocking collective calls to future MPI versions, using CAF and/or UPC to perform the operation is currently the only way to do this. In the following simple example, a CAF global sum is shown. With this infrastructure many areas are shown where the individual processors can perform other operations.

To implement the global sum, we will use a simple one-level tree, where we designate a master task and the child tasks. All tasks will perform their local sums and transfer their sums to the master. The master spins on a receive loop until all the sums arrive from individual children tasks and then performs the global sum and sends the global sum out to the children tasks. While this simple reduction is only doing a sum, the structure can be used for any collective including an MPI_ALLTOALL.

It is important to understand the scoping characteristics of the variables to be used in the global sum. In this example, VOLATILE and SAVE are important to obtain correctness. The SAVE simply saves the value of the variable from one invocation to the next. The VOLATILE forces the compiler to always fetch the operand from memory, rather than getting it from cache.

```
function global_sum_caf (X, MASK)

! -------------------------------------------------------------------
!
! computes the global sum of the _physical domain_ of a 2-d
! array.
!
! -------------------------------------------------------------------

  real (kind=dbl_kind), dimension(imt,jmt), intent(in) ::
&    X,MASK                  ! real multiplicative mask
  real (kind=dbl_kind) ::
&    global_sum_caf          ! resulting global sum
  integer (kind=int_kind) ::
&    i,j                     ! loop indices
  integer (kind=int_kind), save ::
&    me                      ! procid
  logical, save ::
&    first = .true.
  logical, volatile ::
&    children_ready
  logical, volatile, save ::
&    master_ready[*],
&    child_ready(NPROC_X*NPROC_Y)[*]
  real (kind=dbl_kind)::
&    reduce_real_local,
&    global_sum,
&    reduce_real_global(0:NPROC_X*NPROC_Y)[*]
```

It is important that the variable is first scoped with a SAVE. We only want to perform the initialization on the first entry into the routine. It is also important that the variable master_ready and child_ready are scoped VOLATILE. These variables will be placed in the remote processors memory by another processor and we want to make sure that the compiler refetches the variable from memory when it needs the value.

First, all the processors initialize their variables and proceed to compute their local sum.

```
if (first) then
    me = this_image()
    master_ready = .false.
    do i = 1,NPROC_X*NPROC_Y
        child_ready(i) = .false.
    enddo
    first = .false.
    call sync_images()
endif
!
!   sum local contributions
    reduce_real_local = c0
    do j = jphys_b,jphys_e
        do i = iphys_b,iphys_e
            reduce_real_local = reduce_real_local + X(i,j) *
                MASK(i,j)
        end do
    end do
```

Notice the use of this_image() function; this function returns the task number. The sync_images() call is similar to a barrier. The initialization is performed and then the local sum is computed in parallel across all of the tasks. Once the local sums are computed, the children must transfer their sums to the master.

```
!   send local sum to master
    reduce_real_global(me)[1] = reduce_real_local
    call sync_memory()
    child_ready(me)[1] = .true.
```

All processors, including task 1, which is the master, put their sum into the array reduce_real_global(me) in the memory of processor 1. The sync_memory function waits until all transfers from this processor have completed. At this point, the value true is put into the array child_ready(me) on task 1. At this point, all the tasks, except the master waits until the master receives all the sums, performs the global_sum and sends the global_sum back to the children.

```
! sync images replacement
      do while (.not. master_ready)
      enddo
      master_ready = .false.
```

This do_while statement waits until the master task puts true into the variable master_ready into the processor's memory. Once the variable is received from the master, the master_ready is set to false for the next entry into the subroutine. While the children are waiting in the do while loop, it could easily do other operations.

Meanwhile the master is receiving the data from the other processors and performing the global sum.

```
      if(me.eq.1)then
! wait until all local results have arrived
      children_ready = .false.
      do while (.not. children_ready)
        children_ready = .true.
        do i = 2,NPROC_X*NPROC_Y
          children_ready = children_ready .and. child_ready(i)
        enddo
      enddo
      do i = 2,NPROC_X*NPROC_Y
          child_ready(i) = .false.
      enddo
! global sum
      global_sum = reduce_real_global(1)
      do i = 2,NPROC_X*NPROC_Y
          global_sum = global_sum + reduce_real_global(i)
      enddo
! broadcast
      do i = 1,NPROC_X*NPROC_Y
          reduce_real_global(0)[i] = global_sum
      enddo
      call sync_memory()
      do i = 2,NPROC_X*NPROC_Y
          master_ready[i] = .true.
      enddo
```

The first part of the master code waits for the children to place true into the child_ready array. Once all of these values are true, then the master knows that the sums have also been placed. This is a very important point. Note that the child_ready flag is used to indicate that each processor has performed a sync_memory call to assure placement of its sum into the master's memory. Once all of the child_ready flags are true, the master sets all the child_ready flags to false, performs the summation, and places the global_sum and the master_ready flags into the other tasks' memory. At this point, the master is finished and it can perform any other operations. Note that there is no need for a barrier at the end of the global sum. Once a task has the global_sum it can proceed to the next computation or communication.

This is not the most efficient way to perform a global sum with CAF. Note that the master is doing a bulk of the work and all the other tasks are waiting for it to finish its work. An improvement to this sum would be to split up the master's work. As the number of processors grows, one would want to designate teams of processors to perform the sum within the team. For example, it would be good to have a team of processors that represent all of the CAF images (tasks) on a node. This sum could be performed very fast using shared memory accesses. Then each node would proceed through the logic given above. This proposed change would be a two-level reduction tree. If the number of nodes is very large, the user might consider adding another level into the tree to overlap as much of the master work as possible.

Of course, the logic above is much more difficult than calling MPI_ALLREDUCE; however, it does open the door for overlapping the communication required for the operation with some computation that could be performed in parallel. In addition, the example is included here to illustrate how one programs in CAF. If no parallel computation or other communication is available, the user should check to see if the vendor supplies a CAF global sum or just use an MPI_ALLREDUCE call.

4.6 COMPILERS FOR PGAS LANGUAGES

Since CAF and UPC are extensions to Fortran and C, respectively, the compiler must understand the syntax and generate the appropriate get, if the reference is on the right side of the replacement, or a put, if it is on the left side. While some programmers may use straightforward gets and puts such as

$$A(1:N) = B(1:N)[kproc + 1]$$

$$B(1:N)[east\_neighbor] = A(1:N)$$

which is easy for a compiler to convert to a get of N elements of B from processor kproc + 1 and a put of N elements of A in processor B(1:N) on the processor whose value is east_neighbor.

When a CAF syntax is found within a FORTRAN DO loop, it is extremely important that the particular CAF read or write be vectorized out of the DO loop. For example,

```
     do n = 1,recv_num
        pe = recv_pe(n)
        tc = ilenght(recv_length(pe),pe)
        ll = send_low[pe + 1]%p_send_low(mype)
        do l = 1,tc
!dir$ concurrent
        do lll = ilength(l,pe),ilenght(l,pe)-1
           rindex = rsend_index[pe + 1]%p_rsend_index(ll)
           output(recv_index(lll)) = output(recv_index(lll)) + &
               send_scratch[pe + 1]%p_send_scratch(rindex)
           ll = ll + 1
enddo ! lll
enddo ! l
enddo ! n
```

This multinested looping structure from an adaptive mesh refinement code, first loops over the number of neighbors that send data to this processor. Then the second loop loops over the total number of messages that are received from the sending processor. Finally, the inner loop passes over the elements within the message. The logic in the inner DO loop gathers the data from the distant processor and accumulates into the output array. Due to all the indirect addressing, the compiler cannot vectorize the DO loop without help from the user. The !dir$ concurrent directive tells the compiler that it can safely vectorize the code.

With the vectorization of the code, the transfers are performed as a bulk transfer and not element by element. This example shows the power of co-arrays in randomly gathering data from many different processors.

The actual interface to the remote memory accesses is important. Many compilers may translate the CAF or UPC statements to subroutine calls for getting and putting the data. This adds overhead that should not be required. Many of the newer interconnects have low-level interfaces to the interconnect that perform the required operation and these can be accessed directly from the assembly code the compiler generates.

4.7 ROLE OF THE INTERCONNECT

In addition to having a good PGAS compiler, the interconnect must have a good low-level remote memory get and put mechanism that allows the compiler to directly fetch from or store to the remote memory. While there are numerous libraries that allow for gets and puts—SHMEM, GASNET, ARMCI, and so on—these libraries may be built on top of a handshaking message passing library like MPI. If CAF and/or UPC are not built upon an interconnect that has real remote memory gets and puts, the performance will not be good.

The interconnect plays a very important role in the execution of the PGAS language. First, the hardware must support a low-level remote get/put to a distant node. The most effective PGAS programs will have numerous messages in flight in parallel to effectively overlap communication with other communication and/or communication with computation. Regardless of the capability of the compiler, there are numerous cases where the programmer will want to perform a gather or scatter of a large number of small messages, as in single word indirect addressing across the entire parallel system. For this reason, the interconnect must be able to have a large number of outstanding loads and stores from or to a particular node. In addition to the ability to have low-level remote gets and puts, the interconnect must have a very low latency.

PGAS languages give the programmer the ability to directly access a remote processors memory which significantly simplifies parallel programming for the multicore parallel system. Whether the memory being accessed is on the nodes memory or from a remote nodes' memory is transparent in the language. If the PGAS fetches and stores are coming from the nodes memory, the latency will be lower and the bandwidth will be higher. The programmer can easily use MPI for internode communication and PGAS for intranode communication.

EXERCISES

1. How are multiprocessors organized on a node?
2. How do MPI and OpenMP use multiprocessors differently on a node?
3. How does the use of memory differ between MPI and OpenMP?
4. How is MPI communication handled between processors on the same node and processors on different nodes?

5. How do 1D, 2D, and 3D decomposition affect MPI processing? MPI communication? Why select 2D over 1D, or 3D over 2D? Why not?

6. What are the two types of MPI scaling?

7. What are the collective MPI routines and how are they used?

8. How does MPI_ALLTOALL performance change as the number of nodes increases?

9. For what size messages does latency dominate performance?

10. What are the advantages to preposting receives?

11. How is NUMA connected to multisocket nodes?

12. In OpenMP parallelization, what are the two primary variable scoping types?

13. What is false cache line sharing? How does it affect performance?

14. What are the primary methods for partitioning computations among threads?

15. How are mutual exclusion routines used in multithreading?

16. What is PGAS? What compilers support PGAS? What interconnects support PGAS?

17. What are the principal disadvantages of MPI as the multicore sockets grow to larger and larger core counts?

18. A user comes to you with an application that is all-MPI and it needs more memory/core than that available on the node. What might be some actions that he can take to run his code?
    a. Should he try to use fewer cores on the node? Why?
    b. Should he try to use OpenMP on the node? Why?

19. What are some of the advantages of co-array Fortran when writing complex random gather/scatter across the processors in a multicore MPP system?

20. Why is it important that the compiler vectorize a DO loop that contains a CAF get?

21. Does an all MPI application have to be completely rewritten to use CAF?
    a. If the answer is "no" what might a good approach be to incrementally insert CAF into an MPI program.

22. CAF has the notion of teams of processors. How might this be used to address a large multicore MPP system?

23. The global sum performed in this chapter was a single level tree. How might this sum be redone to employ the locality of the cores on a node? Think of having teams of processors on each node.

CHAPTER 5

A Strategy for Porting an Application to a Large MPP System

O PTIMIZING AN APPLICATION to run across tens of thousands of processors is a very difficult task and should be performed in a methodical manner. The most important issue to be addressed, when optimizing an application, is to gather runtime statistics. Do not ever believe someone who says "Oh, all the time is spent in CRUNCH." Oftentimes, the statistics are surprising, especially when running the application at scale. Also make sure that a typical production problem is obtained. Toy problems, to illustrate that an application runs correctly, are not what is needed. Also consider a problem that needs large processor counts to solve a breakthrough science problem. Ideally, the targeted problem to be solved is much larger than the one currently being addressed on smaller systems. If the target is to make large science studies with an application, then the input to be used when collecting statistics is the input for the large science study.

When moving to large processor counts, two scaling approaches may be used. Strong scaling is the case where the problem to be solved is set, and as the number of processors increases, the amount of work performed by each processor decreases proportionally. On the other hand, weak scaling sets the amount of work per processor constant, and as the number of processors increases, the amount of total work increases as well. Weak scaling is easier to scale, since the amount of computation performed by each processor is constant and that help hide the communication. In weak scaling cases, the increased work is typically in the form of a finer or larger grid in a finite-difference application or more particles in a particle in cell (PIC) application. It is important to only increase the work in the problem when it benefits the science being delivered. If the increased work is not beneficial to the scsience, then one must stay with the problem size that delivers the maximum understanding of the problem being solved and proceed in a strong scaling mode.

Typically, the application should be run until the time-step integration makes up a majority of the runtime. The initialization phase is not of interest, only the major computational routines in the analysis phase of the application, since in most production runs, the initialization phase is not important and can be ignored. An exception to this is when a complex decomposition scheme is used to decompose the problem. In this case, the decomposition can grow exponentially as the number of processors increase as with the Leslie 3D application in a later chapter. If this is the situation, more efficient and/or parallel decomposition schemes should be employed.

Once the problem definition is obtained, a scaling study should be performed using a production data set. The scaling study should consist of four to five runs from thousands of processors to tens of thousands of processors or even higher.

When examining the results, look for the first deviation from perfect scaling and gather profiling information to determine the cause of the deviation. The cause of the deviation can be identified by gathering statistics for those few runs prior to and after the deviation from perfect scaling. Computation, communication, and I/O statistics should be gathered. Numerous tools are available to gather these statistics. For node performance, one wants to have the typical "gprof" profile data with more specific information as to what line numbers within the routine are taking the time which is extremely important when the most compute-intensive subroutine contains a large number of lines of code.

When looking at message passing statistics, load imbalance is the first set of data that needs to be examined. Load balance can come from either computation imbalance or communication imbalance. When load imbalance exists, the first task is to determine its cause. Do not assume that load imbalance is caused by communication; although the load imbalance might show itself within the first MPI operation after the load imbalance occurs. For example, when an algorithm requires a global sum, all the processors must arrive at the global sum prior to performing the operation. Many performance analysis tools contribute this wait time to the global sum when it is due to load imbalance. When gathering statistics, it is always good to gather communication statistics in such a way that the required barrier prior to a collective is measured separately from the collective itself. The inclusion of the barrier prior to the global reduction gives the time the processors need to synchronize and it shows when the processors become out of balance. Communication statistics should include times for all types of message passing with information as to the message sizes and time consumed by each MPI operation.

Once statistics are in hand, the first bottleneck needs to be examined. If the computation time is greater than 50% of the wall-clock time, node optimization should be examined first. If, on the other hand, communication uses most of the time, it should be examined first. It is very important not to spend time optimizing a particular computation routine that uses most of the time on a few processors. Oftentimes, scaling to a higher number of processors may reduce its time so that it is no longer the bottleneck. Other routines that do not scale may become the bottleneck at the higher processor counts. Do not waste time optimizing something that is not important at the processor counts where production would be running. The following analysis uses the CrayPat™ tool from Cray Inc. Its capabilities are not unique; other tools or collection of tools can give similar data.

Consider the following scaling chart for the S3D application. S3D is a three-dimensional combustion code that uses a high-order differencing scheme. Figure 5.1 gives the scaling for a "weak scaling" problem set. Once again, as the number of processors is increased, the work increases accordingly hence, the total runtime across the range of processors should be constant.

Note that we get reasonable scaling from 1500 to 12,000 processors, and then at 48,000 processors, we get very poor scaling. Given these data, we might suspect that communication or I/O might be the bottleneck. Given the weak scaling characteristic of the study, the computation would never cause a factor of two deviations at high processor counts. We will investigate the reason for this performance degradation in Chapter 7.

5.1 GATHERING STATISTICS FOR A LARGE PARALLEL PROGRAM

The larger and more complex the application, the more difficult it is to gather statistics. There are numerous statistics gathering approaches. First is the traditional profiling approach. Here, either a preprocessor or a compiler instruments the binary to gather summary information on the execution of each and every subroutine and function that is called during the execution of the program. The typical default for profiling tools is to instrument everything. The problem with this approach is that the instrumentation can introduce overhead and skew results for very small subroutines and functions. Additionally, an application with a very large number of small routine calls may run significantly longer when instrumented.

A second approach is to sample the executing program. Sampling usually will look at the program pointer every few microseconds. This sampling approach does not introduce as much overhead and it also has the benefit that it can gather information on finer boundaries than routine boundaries. With sampling it is possible to get line-level execution profiling without additional overhead. The disadvantage of this approach is that one does not have the detailed information available to the traditional profiling discussed above.

Developers at Cray Inc. came up with an interesting combination of the two approaches which gives one the best from both approaches [11]. Unfortunately, one has to run your application twice, first using the sampling approach and then using a filtered profiling approach. The first run of the application uses sampling to identify the most important routines in the applications. Additionally with the sampling, the small routines that use a miniscule amount of time each time they are called can be identified and avoided. This first approach gives an overall summary of the execution of the application. Consider the following sampling profile obtained from running a combustion application on 6000 processors. Note that the time is broken into three sections: MPI time, User time, and time taken by miscellaneous system routines (ETC). The MPI is then broken into MPI routines that take up a majority of the time; in this case, MPI_WAIT and MPI_WAITALL. Next, the User time, which accounts for 28.9% of the overall runtime, is broken down into the individual routines that comprise the time. Finally, ETC gives library routines and the time they utilize. The default is that no element is displayed that utilizes less than 1% of the time. The first column in Table 5.1 gives the percentage of the samples, the second gives the number of samples, the third gives the imbalance of the sampling across all 6000 processors, and finally, the fourth column gives the imbalance percentage. This table shows one of the weaknesses of sampling. There is no way of knowing which routine calls some of the important library calls.

TABLE 5.1 Profile by Function

Samp %	Samp	Imb. Samp	Imb. Samp %	Experiment=1 Group Function PE = 'HIDE'
100.0%	89319	--	--	Total
46.8%	41827	--	--	MPI
43.8%	39138	4328.78	10.0%	mpi_wait_
1.0%	864	3287.25	79.2%	mpi_waitall_
35.5%	31720	--	--	USER
5.1%	4560	380.49	7.7%	mcavis_new_looptool_
4.1%	3632	1000.14	21.6%	rhsf_
2.8%	2499	203.94	7.5%	diffflux_proc_looptool_
2.3%	2070	162.47	7.3%	rdot_
1.9%	1721	173.56	9.2%	ratx_
1.8%	1617	392.03	19.5%	integrate_erk_jstage_lt_
1.7%	1534	161.81	9.5%	rdsmh_
1.5%	1333	325.02	19.6%	computeheatflux_looptool_
1.4%	1286	122.76	8.7%	ratt_
1.4%	1237	158.32	11.4%	chemkin_m_reaction_rate_bounds_
1.0%	883	105.70	10.7%	qssa_
1.0%	870	313.07	26.5%	thermchem_m_calc_inv_avg_mol_wt_
17.7%	15772	--	--	ETC
11.6%	10347	416.23	3.9%	__fmth_i_dexp_gh
1.5%	1360	142.68	9.5%	__fvdexp_gh
1.4%	1217	277.71	18.6%	__c_mcopy8
1.0%	858	121.15	12.4%	__fmth_i_dlog10_gh

Several important facts can be drawn from this table. At this processor count, this application is slightly computationally dominant. MPI is utilizing less than 50% of the time. When we delve into optimizing this application, we first look at how to improve the computation. There is also significant load imbalance in some of the computational routines. A second bit of information we can obtain from sampling is the actual line number in the source code that is responsible for the maximum amount of time. Table 5.2 gives us some details for this run.

Note that under the subroutine mcavis_new_looptool that uses 4.2% of the compute time, line number 103 uses 4% of the compute time. This shows that the Fortran statement that starts at line number 103 uses all but 0.2% of the entire routine. Undoubtedly, line number 103 points to the start of a DO loop. In addition to this useful information, we can now use our profiling approach where we only instrument the top-level routines as

TABLE 5.2   Major Computational Routines

| 28.9%	|76550	| --	| --	|USER
||---				
|| 4.2%	|11238	| --	| --	|mcavis_new_looptool_
3|	|	|	|	|source/f77_files/johnmc/mcavis_new_lt_gen.f
||||---				
4|||4.0%	|10616	|1010.30	|8.7%	|line.103
||||===				
|| 3.4%	| 8983	| --	| --	|rhsf_
3|	|	|	|	|source/f90_files/solve/rhsf.f90
|| 2.0%	| 5185	| 259.74	|4.8%	|rdot_
3|	|	|	|	|f77_files/C7H16_52species/std/getrates.f
||||---				
4|||	|	|	|	|line.2628
||||===				
|| 2.0%	| 5183	| --	| --	|diffflux_proc_looptool_
3|	|	|	|	|source/f77_files/johnmc/diffflux_gen_uj.f
||||---				
4|||1.9%	| 5161	| 521.01	|9.2%	|line.199
||||===				

well as all of the MPI libraries. From a second run of the application with the profiling, we will be able to extract more pertinent information.

In Table 5.3, we only get the routines we have profiled and the main program. All routines that are called from within these listed routines which are not profiled are added to the time in the profiled routine. For example, getrates was one of the major routines that called dexp; so dexp's time is included into the time for getrates. The numbers in the second and third columns of the table are in seconds. Once again, the fourth column contains the load imbalance percentage.

Table 5.4 gives us the MPI statistics from the profile run. This table shows us the major MPI routines and where they are called.

TABLE 5.3   Profile by Function Group and Function

Time %	Time	Imb. Time	Imb. Time %	Calls	Experiment = 1 Group Function PE = 'HIDE'
100.0%	1530.892958	--	--	27414118.0	Total
52.0%	796.046937	--	--	22403802.0	USER
22.3%	341.176468	3.482338	1.0%	19200000.0	getrates_
17.4%	266.542501	35.451437	11.7%	1200.0	rhsf_
5.1%	78.772615	0.532703	0.7%	3200000.0	mcavis_new_looptool_
2.6%	40.477488	2.889609	6.7%	1200.0	diffflux_proc_looptool_
2.1%	31.666938	6.785575	17.6%	200.0	integrate_erk_jstage_lt_
1.4%	21.318895	5.042270	19.1%	1200.0	computeheatflux_looptool_
1.1%	16.091956	6.863891	29.9%	1.0	main
47.4%	725.049709	--	--	5006632.0	MPI
43.8%	670.742304	83.143600	11.0%	2389440.0	mpi_wait_
1.9%	28.821882	281.694997	90.7%	1284320.0	mpi_isend_

TABLE 5.4   MPI Message Stats by Caller

MPI Msg Bytes	MPI Msg Count	MsgSz <16B Count	16B <= MsgSz <256B Count	256B <= MsgSz <4KB Count	4KB <= MsgSz <64KB Count	64KB <= MsgSz <1MB Count	Experiment = 1 Function Caller PE [mmm]
25634067336.0	1287990.0	3596.0	65.0	89606.0	1194722.0	1.0	Total
25633587200.0	1284320.0	--	--	89600.0	1194720.0	--	mpi_isend_
6635520000.0	259200.0	--	--	--	259200.0	--	derivative_z_send_
3							rhsf_
4							integrate_erk_
5							integrate_
6							solve_driver_
7							MAIN_
8	6635520000.0	259200.0	--	--	--	259200.0	--	pe.1801
8	6635520000.0	259200.0	--	--	--	259200.0	--	pe.4663
8	6635520000.0	259200.0	--	--	--	259200.0	--	pe.3666

The order of the MPI routines is determined by the calls that result in maximum communication. Note that MPI_ISEND is the routine responsible for sending most of the data. This explains somewhat why all the time is spent in MPI_WAIT. This is a nonblocking MPI operation; in other words, the message is sent and the program continues computing while the message is transferred. To find out when the message is completed, the program must call MPI_WAIT for a particular message or MPI_WAIT_ALL for a group of messages. Additionally, we see that 6,635,520,000 out of 25,633,587,200 messages are sent from derivative_z_send. The table continues to include the other routines that call MPI_ISEND. If we look into derivative_z_send, we will probably be able to find the MPI_WAIT calls.

In addition to the MPI information, we also have hardware counter information from this execution. Table 5.5 gives hardware counter data for getrates. Note that there are four PAPI counters that are used. Unfortunately, on most systems there are a limited number of hardware counters that can be used at any given time. The four are L1 data cache misses, TLB data misses, L1 data cache accesses, and floating point operations. From these four raw metrics the software computes some derived metrics for us: computational intensity, MFLOP rate, TLB, and L1 references per miss. Additionally, we see that this routine achieves 9.7% of the peak. These statistics are summed over all 6000 processors and the averages are being displayed. From these numbers we see that the TLB

TABLE 5.5   Hardware Counters for Major Routines

```
================================================================================
==
USER/getrates _
--------------------------------------------------------------------------------
--
Time%                                             22.3%
Time                                              341.176468 secs
Imb.Time                                          3.482338 secs
Imb.Time%                                         1.0%
Calls                          0.055 M/sec        19200000.0 calls
PAPI _ L1 _ DCM                6.291 M/sec        2191246517 misses
PAPI _ TLB _ DM                0.024 M/sec        8250904 misses
PAPI _ L1 _ DCA                1044.783 M/sec     363928615376 refs
PAPI _ FP _ OPS                1008.976 M/sec     351456000000 ops
User time (approx)             348.329 secs       905656290778 cycles  100.0%Time
Average Time per Call                             0.000018 sec
CrayPat Overhead : Time        3.8%
HW FP Ops/User time            1008.976 M/sec     3 5 1 4 5 6 0 0 0 0 0 0   o p s
  9.7%peak(DP)
HW FP Ops/WCT                  1008.976 M/sec
Computational intensity        0.39 ops/cycle     0.97 ops/ref
MFLOPS (aggregate)             6053856.92 M/sec
TLB utilization                44107.73 refs/miss 86.148 avg uses
D1 cache hit,miss ratios       99.4% hits         0.6% misses
D1 cache utilization (misses)  166.08 refs/miss   20.760 avg hits
```

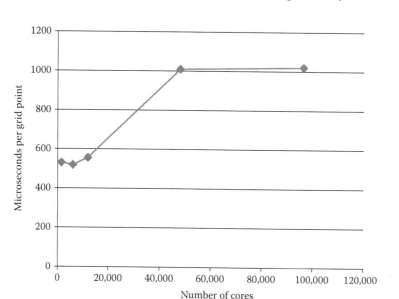

FIGURE 5.1   Weak scaling study of S3D.

utilization is excellent and L1 cache reuse is very good. The metric that might be a little low is the computational intensity. This indicates that there are approximately an even number of memory access and floating point operations. One would like this to be somewhat higher; however, it is not always possible to increase the computational intensity. From this information, we have an excellent place to start strategizing an approach to improve the performance of this application.

The profile data presented above were from the S3D run in Figure 5.1 at 6000 processors. The following profile is from the S3D run at 48,000 processors.

We see a completely different story with these data. Clearly, communication is now taking 76.9% of the time. We will investigate how to address the optimization of S3D at 6000 cores in Chapter 6 on core optimization and the 48,000 core problem in Chapter 7 on communication. If we want to run our production at 6000 cores, computation is the bottleneck, and if, on the other hand, we want to run our production at 48,000 cores, we better do something about the communication.

In this case, we assume that the massively parallel processor (MPP) system is not at fault in causing the bottleneck. In some cases it could be that the design of the parallel system is not meant to handle processor runs greater than 5000–10,000.

TABLE 5.6   Profile by Function Group and Function

Time %	Time	Imb. Time	Imb. Time %	Calls	Experiment=1 Group Function PE = 'HIDE'
100.0%	1730.555208	--	--	16090113.8	Total
76.9%	1330.111350	--	--	4882627.8	MPI
72.1%	1247.436960	54.277263	4.2%	2389440.0	mpi_wait_
1.3%	22.712017	101.212360	81.7%	1234718.3	mpi_isend_
1.0%	17.623757	4.642004	20.9%	1.0	mpi_comm_dup_
1.0%	16.849281	71.805979	81.0%	1234718.3	mpi_irecv_
1.0%	16.835691	192.820387	92.0%	19999.2	mpi_waitall_
22.2%	384.978417	--	--	11203802.0	USER
9.9%	171.440025	1.929439	1.1%	9600000.0	getrates_
7.7%	133.599580	19.572807	12.8%	1200.0	rhsf_
2.3%	39.465572	0.600168	1.5%	1600000.0	mcavis_new_looptool_

In these profiles we did not see any I/O show-up; however, that is not always the case. In fact in most cases of applications that are being moved to 100,000s of processors, the I/O becomes the critical issue.

Now the following chapters are organized to address each of the potential bottlenecks.

a. Computation or user time is using a majority of the time (Chapter 6).

b. Communication is using a majority of the time (Chapter 7).

c. I/O is using a majority of the time (Chapter 7).

So where does OpenMP™ come in? In Chapter 7, we will see that oftentimes the solution to improving an application that does not scale well is to introduce OpenMP. Chapter 8 discusses the issues of using OpenMP efficiently.

EXERCISES

1. What is the first task in porting a program to a large MPP system?
   a. What are some of the reasons for moving a scientific computation to larger and larger processor counts?
2. What kind of problem should be used?
   a. Why should long runs be made when gathering statistics?

    b. If a statistics gathering package had the ability to turn the profiling on and off, how could this be used to address the time spent in the initialization?

3. What are the two types of scaling? How do they differ?

    a. What are some of the ways to increase work when using weak scaling?

4. How does load balance/imbalance affect performance?

    a. How does load imbalance usually show up?

5. What are possible problems when running an instrumented code?

6. Why is it important to look at profiles at large processors counts when one already has a profile at a small processor count run?

CHAPTER 6

Single Core Optimization

I N CHAPTER 5, THE STRATEGY for approaching the optimization of a large application for an MPP is given. If the majority of the runtime indicates that the compute time is the bottleneck, this is the right chapter to look at. We look at some simple but informative kernels that illustrate the right and wrong way to program for multicore processors. In this chapter we are only concerned with optimizing on a single core and cover parallelization over multiple cores later.

6.1 MEMORY ACCESSING

In Chapter 1, the table look-aside buffer (TLB) was discussed. Efficient utilization of the TLB is very important for efficient CPU code and the first potential bottleneck to examine. Consider the following looping structure:

```
DO I = 1,100
DO J = 1,100
DO K = 1,100
    A(I,J,K) = B(I,J,K) + C(I,J,K)
ENDDO
ENDDO
ENDDO
```

Assume that the dimensions of the arrays A, B, and C are 100, 100, and 100 and that they are REAL(8) operands. The first time through the DO loop each of the array elements A(1,1,K), B(1,1,K), and C(1,1,K) that are fetched will be on different physical pages since the size of each array

(10,000 words or 80,000 bytes) is larger than a page, and for each invocation of the inner DO loop, 300 different pages will be accessed. The loop basically fetches planes and only uses one point on each plane in the DO loop. The problem is that only 48 unique pages can be held within the TLB and on each invocation of the inner DO loop all 300 pages will be continuously reloaded into the TLB. Every access of the operands within this looping structure incurs a TLB miss and the memory access time is twice what it could be.

The following DO loop structure is better:

```
DO K=1,100
DO J=1,100
DO I=1,100
   A(I,J,K)=B(I,J,K)+C(I,J,K)
ENDDO
ENDDO
ENDDO
```

Only three pages are loaded for each invocation of the inner DO loop. As the J loop increments, additional pages may be referenced; however, every element of the page is utilized. Hence, the TLB is well used. Timing these two loops on a system would probably show identical times, since any smart compiler will automatically bring the I loop inside in this simple example; however, the user should check the comments put out by the compiler to assure that it reorders the DO loops. Caution—on more complex looping structures, determining the correctness of performing this loop inversion may be too complex for a compiler to perform and if the compiler cannot invert the loops, the performance difference between the two looping structures would be significant.

In the following examples, the Opteron™ core was running at 2.1 GHZ. The Opteron can generate four 64-bit results/clockcycle when the DO loop is vectorized. The peak performance for one core is 8.4 GFlops. In these examples, unless otherwise noted, the Cray® compiler is used. In Figure 6.1, the performance is plotted as a function of computational intensity with a vector length of 461.

The computational intensity is the number of operations in the loop divided by the number of fetches and stores. As the computational intensity increases, the reliance on memory bandwidth decreases. The higher level of computational intensity comes from high-order polynomials and is representative of log, exp, power, and so on.

Example Loop	Computational Intensity	Bottleneck
A(:) = B(:) + C(:)	0.333	Memory
A(:) = C0 * B(:)	0.5	Memory
A(:) = B(:) * C(:) + D(:)	0.5	Memory
A(:) = B(:) * C(:) + D(:) * E(:)	0.6	Memory and multiply
A(:) = C0 * B(:) + C(:)	0.667	Memory
A(:) = C0 + B(:) * C1	1.0	Still memory
A(:) = C0 + B(:) * (C1 + B(:) * C2)	2.0	Add and multiply

In this figure, we see an increased performance as the computational intensity increases. In later sections, techniques for increasing the computational intensity are discussed.

The performance for the indirect addressing is less than the contiguous case, even when the computational intensity is large enough to amortize the memory bandwidth. The reason for this difference is the nonvectorization of the DO loops that are accessing the arrays indirectly or with a stride. Since the compiler does not vectorize DO loops containing indirect addressing, the indirect addressing examples as well as the STRIDE = 128 examples show scalar performance in every case. The computational intensity in the indirect addressing examples does not consider the

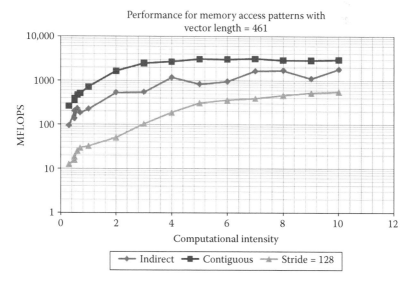

FIGURE 6.1   Performance of kernels as a function of computational intensity.

fetching of the address array; hence, technically there are more fetches than the contiguous case. For example,

```
DO I = 1,N
   A(IA(I)) = B(IB(I)) + C(IC(I))
ENDDO
```

While this computational intensity is plotted at 1/3, it is actually 1/6.

6.1.1 Computational Intensity

Computational intensity is one of the important characteristics of an application. Today we see some applications that are able to achieve 60–70% of the peak on a large MPP system while others achieve only 5–10% of the peak. The difference between these applications is computational intensity of the major compute kernels. The applications that achieve less than 10% of peak are undoubtedly memory bandwidth limited while the applications that are able to achieve very high percentage of the peak are typically dependent on matrix multiply. For example, the HPL benchmark, used to determine the top 500 list, is heavily dependent on matrix, matrix multiply. On the top 500 list, the percentages of peak that the large system achieve is somewhere between 70% and 80% of peak. When the matrix multiply is coded for an optimal library, the computation intensity can approach 2. For example, consider the following naïve coding of the matrix multiply loop.

```
      DO 46030 J = 1, N
       DO 46030 I = 1, N
         A(I,J) = 0.
46030 CONTINUE
      DO 46031 K = 1, N
       DO 46031 J = 1, N
         DO 46031 I = 1, N
           A(I,J) = A(I,J) + B(I,K) * C(K,J)
46031 CONTINUE
```

What is the computational intensity of the inner DO 46031 loop? $C(K,J)$ is a scalar with respect to the loop, so we have two memory loads, $A(I,J)$ and $B(I,K)$, one memory store $A(I,J)$ and two floating point operations. The computational intensity is 2/3 (Floating point operations/ Number of memory operations). Is there any way that the computational intensity of this DO loop can be improved? Consider the following rewrite:

```
      DO 46032   J = 1, N
        DO 46032 I = 1, N
         A(I,J)=0.
46032 CONTINUE
C
      DO 46033 K = 1, N-5, 6
        DO 46033 J = 1, N
          DO 46033 I = 1, N
           A(I,J) = A(I,J) + B(I,K)   *  C(K,J)
     *                     + B(I,K+1)  *  C(K+1,J)
     *                     + B(I,K+2)  *  C(K+2,J)
     *                     + B(I,K+3)  *  C(K+3,J)
     *                     + B(I,K+4)  *  C(K+4,J)
     *                     + B(I,K+5)  *  C(K+5,J)
46033 CONTINUE
C
```

In this example, we unroll the K loop inside the J loop. Why the K loop? Note that the original loop had only one array reference on K, remember C(K,J) is a scalar with respect to the I loop, and the J loop has two array references on J. When we unroll the K loop, we introduce one additional fetch and two floating point operations with each unroll. Now look at the computational intensity of the inner DO loop. All the Cs are scalars; we have seven memory loads, all the Bs and A(I,J) and one store, and we have 12 floating point operations. The computational intensity of the rewritten DO loop is $12/8 = 1.5$. Figure 6.2 shows how this impacts the compiler ability to generate an efficient code.

Whenever an application performs a matrix product, it may be best to insert a call to the appropriate library routine; however, this is a good example of how the computational intensity of an application might be improved. Note in this example that we also compare a C version of the routine versus the Fortran version. In both cases, the original and the restructured, the Fortran code outperforms the C code. Most compilers are more adapt at optimizing Fortran, since there are more potential side-effects in C.

6.1.2 Looking at Poor Memory Utilization

The next example looks at striding through memory versus contiguous accessing of the arrays. This example is run in two different modes. Most

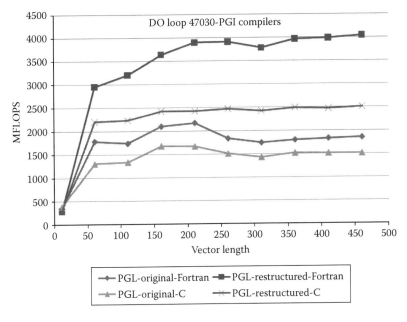

FIGURE 6.2 Performance of original and restructured matrix multiply in Fortran and C.

examples in this chapter are run in the unpacked mode. That is, it is run on one of the cores of the multicore socket. The performance that one obtains from such runs is the best case. The reason for this is that all the bandwidth available to the socket are dedicated to the core. A second way to run the example is in the packed mode. In this mode, the kernel is run on every core of a socket at the same time. This is more common when running an OpenMP™ and MPI application. When running in packed mode, there is more impact on the memory bandwidth.

```
DO 41090 K = KA,  KE,  -1
  DO 41090 J = JA,  JE
    DO 41090 I = IA,  IE
      A(K,L,I,J) = A(K,L,I,J) - B(J,1,i,k) * A(K+1,L,I,1)
 *       - B(J,2,i,k) * A(K+1,L,I,2) - B(J,3,i,k) * A(K+1,L,I,3)
 *       - B(J,4,i,k) * A(K+1,L,I,4) - B(J,5,i,k) * A(K+1,L,I,5)
41090 CONTINUE
```

This kernel has also been written in C which follows:

```
// THE ORIGINAL
  for(K=KA; K>=KE; K--){
```

```
for(J=JA; J<=JE; J++){
for(I=IA; I<=IE; I++){
    A[J][I][L][K] = A[J][I][L][K] - B[K][I][1][J]*A[1][I][L][K+1]
      - B[K][I][2][J]*A[2][I][L][K+1] - B[K][I][3][J]*A[3][I][L][K+1]
      - B[K][I][4][J]*A[4][I][L][K+1] - B[K][I][5][J]*A[5][I][L][K+1];
}
}
}
```

This loop nest is extremely inefficient. The stride on the arrays accessed by the inner DO loop is the product of the first and second dimensions of the Fortran array and the last two dimensions in the C code. In this example, both the A array and the B array were dimensioned (8,8,500,8) in the Fortran code and A[8][IIDIM][8][8] in the C code. This example is striding by 64 on each of the arrays. Remember in C, the contiguous accessing is on the othermost index. This will result in poor TLB and cache utilization. One possible restructuring would be to reorganize the arrays as follows:

In Fortran:

```
DO 41091 K = KA, KE, -1
  DO 41091 J = JA, JE
    DO 41091 I = IA, IE
      AA(I,K,L,J) = AA(I,K,L,J) - BB(I,J,1,K) * AA(I,K+1,L,1)
*       - BB(I,J,2,K) * AA(I,K+1,L,2) - BB(I,J,3,K) * AA(I,K+1,L,3)
*       - BB(I,J,4,K) * AA(I,K+1,L,4) - BB(I,J,5,K) * AA(I,K+1,L,5)
41091 CONTINUE
```

And in C:

```
// THE RESTRUCTURED
  for(K = KA;  K >=KE;  K-){
  for(J = JA;  J <=JE;  J++){
  for(I = IA;  I <=IE;  I++){
    AA[J][L][K][I] = AA[J][L][K][I] - BB[K][1][J][I] * AA[1][L][K+1][I]
    - BB[K][2][J][I] * AA[2][L][K+1][I] - BB[K][3][J][I] * AA[3][L][K+1][I]
    - BB[K][4][J][I] * AA[4][L][K+1][I] - BB[K][5][J][I] * AA[5][L][K+1][I];
}
}
}
```

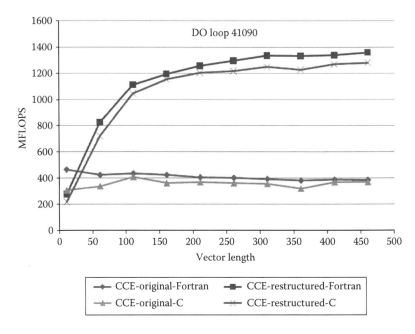

FIGURE 6.3   Performance of original and restructured DO loop 41090 in Fortran and C, unpacked mode.

In this restructuring, the arrays are reorganized to allow contiguous accessing on the inner DO loop. Since just this simple kernel is shown, it is not clear how far-reaching the effects of this restructuring are. If the modified arrays are passed through a module, common block or as subroutine arguments, then the arrays must be reorganized in the other routines that use them. The plot in Figure 6.3 illustrates the performance of this loop and its restructuring that reorganizes the data structures. This chart and the following statistics are from running in the unpacked mode; that is, running only on the core of the socket.

Once again, the performance is very poor for the original, never achieving over 600 MFLOPS. In this example, the performance of the Fortran and C code is very similar with Fortran being a little faster. Striding through arrays hurts TLB performance as well as cache performance. To illustrate this point, the following hardware counters were gathered during the execution of the original:

```
Time%                                    0.0%
Time                                     0.001883 secs
Calls                          --             10.0 calls
```

```
PAPI_L1_DCM                         --               97468 misses
PAPI_TLB_DM                         --                3716 misses
PAPI_L1_DCA                         --              481278 refs
PAPI_FP_OPS                         --              472000 ops
User time (approx)                               0 cycles 0.0%Time
Average Time per Call       0.000188 sec
CrayPat Overhead : Time     0.7%
HW FP Ops / User time               --              472000 ops
HW FP Ops / WCT                     --
Computational intensity     -- ops/cycle          0.98 ops/ref
MFLOPS (aggregate)          --M/sec
TLB utilization             129.52 refs/miss      0.253 avg uses
D1 cache hit, miss ratios   79.7% hits            20.3% misses
D1 cache utilization (M)    4.94 refs/miss        0.617 avg uses
```

On the system being investigated in this book, only four hardware counters can be measured on a given instantiation of the executable. In this case, the four counters are illustrated by the lines starting with PAPI–. The derived metrics follow these lines and give the important statistics. First we see that there are only 129.52 average memory references per TLB miss. This is very poor. Since we are working on 8-byte operands, that number would be 512 if each operand accessible within the page were referenced. Next, when each cache line is loaded into Level 1 cache, only 4.94 uses are made out of a total of 8 available. The striding in the original code hurts both TLB and Level 1 cache usage. Following are statistics gathered from the restructured version of the loop:

```
Time%                                               0.0%
Time                                     0.000626 secs
Calls                           --          10.0 calls
PAPI_L1_DCM                     --         12099 misses
PAPI_TLB_DM                     --          1419 misses
PAPI_L1_DCA                     --        418193 refs
PAPI_FP_OPS                     --        472000 ops
User time (approx)                    0 cycles 0.0%Time
Average Time per Call            0.000063 sec
CrayPat Overhead : Time     2.0%
HW FP Ops / User time           --        472000 ops
HW FP Ops / WCT                 --
Computational intensity         --    ops/cycle 1.13 ops/ref
MFLOPS (aggregate)          -M/sec
TLB utilization             294.71 refs/miss    0.576 avg uses
```

```
D1 cache hit,miss ratios    97.1% hits         2.9% misses
D1 cache utilization (M)     34.56 refs/miss    4.321 avg uses
```

In these statistics we see that the TLB usage has doubled, still not as high it could be. The Level 1 cache utilization has increased by a factor of 8, we are now getting cache reuse rather than not using all the elements in the cache line, we obtain an average of 4.321 uses of each operand. Given this improvement, we see a definite increase in the performance.

This example was run with only one core on the socket active and therefore that core was able to utilize the entire memory bandwidth into the socket. A more conservative comparison is when all the cores of the socket are used. This is achieved by running the kernel in an MPI program with an MPI_BARRIER before and after the DO loops. These results show a marked degradation in the performance of the system due to the contention for the memory bandwidth.

Also, while we are still getting an improvement over the restructured, the overall performance is less than half of the single core run. Users should be aware that timing individual kernels should be performed in this packed mode; that is, all the cores accessing memory at the same time. The resultant performance would be more applicable to an MPI program. This timing is somewhat conservative, since synchronization is forced by the MPI_BARRIERS. In a typical MPI program, the separate MPI tasks are free to operate out of lock step and can therefore spread the memory utilization over a wider period of time. While the first timing given for this DO loop is optimistic and represents performance higher than what would be expected from an MPI program containing this kernel, the second timing is lower than what would be achieved from the MPI program.

In the following chart, we compare the performance of a packed and an unpacked mode. The packed mode is when the example is run on each of the four cores of the socket at the same time and unpacked is when only one core of the socket is used and the other cores are inactive. In the first case, memory is stressed, while in the second case, we strictly test the processor performance since all of the nodes' memory bandwidth is dedicated to a single core (Figure 6.4).

In summary, those loops that are memory limited, owing to small computational intensity and/or striding or indirect addressing, see a larger impact from running in packed mode. Even when the loop is restructured to enable vectorization, if the computational intensity is small, the impact from the packed runs will be significant. One way of reducing the

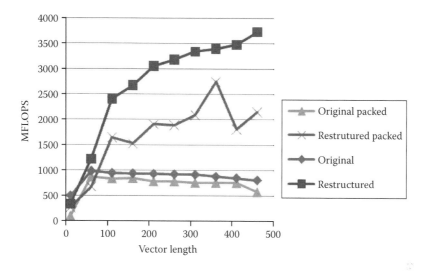

FIGURE 6.4   Comparison of packed and unpacked execution of the original and restructured Fortran DO loop.

dependence on memory bandwidth is to do cache blocking and/or memory prefetching to decrease memory wait time. These techniques are addressed in later chapters.

6.2 VECTORIZATION

In this context, we refer to vectorization as the ability of the target compiler to translate a portion of the application to utilize the SSE3 instructions. These instructions include

1. Memory load and store

2. Memory moves and copies

3. Floating point add

4. Floating point multiply

When the SSE3 instructions are used, each operation produces two 64-bit or four 32-bit results in each clock cycle. Technically speaking, the vector length is only two; however, you will not get good performance if all of your DO loops are of length two. Generally, Fortran loops with vector lengths of 20–30 will give the compiler enough work to amortize the overhead needed to set up the SSE instructions.

6.2.1 Memory Accesses in Vectorized Code

SSE3 instructions require that the data must be contiguous in a 128-bit register. So the compiler must deal with a Fortran DO loop with a non-unitary stride or one that accesses an indirectly addressed array. One method it uses is to issue the appropriate instructions to gather the non-contiguous elements into contiguous registers and then scatter the result registers back into the cache line(s). All compilers can do this, but the overhead introduced from the added data movement in and out of the cache lines can degrade performance. Accessing the arrays in the loop contiguously will promote good vectorization opportunities for the compiler. Here is an example of contiguous accessing in a loop that will vectorize:

```
DO 41033 I = 1, N
  Y(I) = c0 + X(I) * (C1 + X(I) * (C2 + X(I) * (C3 + X(I))))
  41033 CONTINUE
```

But the next loop indexes the array indirectly through another array, and would require gather/scatter instructions to make vectorization possible. Most compilers would not vectorize this loop because of the overhead:

```
DO 41012 I = 1, N
  Y(IY(I)) = c0 + X(IX(I)) * (C1 + X(IX(I)) * (C2 + X(IX(I))))
41012  CONTINUE
```

Indexing arrays with a stride, as in the following loop, would also require gather/scatter operations, and so the loop would probably not be vectorized:

```
    II = 1
  DO 41072 I = 1, N
    Y(II) = c0 + X(II) * (C1 + X(II) * (C2 + X(II)))
    II = II + ISTRIDE
41072 CONTINUE
```

Unlike the hardware on legacy vector machines, which would have achieved good performance on these loops, the SSE vector instructions on current x86 processors are not powerful enough to avoid performance losses due to data motion when indirect addressing or strides appear in loops. A good example of a restructuring to achieve vectorization that was used extensively on the legacy vector machines is the rewrite of a recursive sweep routine. For example, the following DO loop does not vectorize, because it is recursive.

```
      DO 43140 J = 2, N
        DO 43140 I = 2, N
          A(I,J,1) = A(I,J,1)  - B(I,J) * A(I-1,J,1)
    *                          - C(I,J) * A(I,J-1,1)
          A(I,J,2) = A(I,J,2)  - B(I,J) * A(I-1,J,2)
    *                          - C(I,J) * A(I,J-1,2)
          A(I,J,3) = A(I,J,3)  - B(I,J) * A(I-1,J,3)
    *                          - C(I,J) * A(I,J-1,3)
43140 CONTINUE
```

We cannot vectorize on the I index because of the data dependency on the A(I − 1,J) arrays. Also we cannot vectorize on the J index because of the data dependency on the A(I,J − 1). The method we employ is to vectorize the diagonals of the plane. This is somewhat of a complicated rewrite; the important thing is that we have to introduce a stride that would hurt the performance on the SSE3 instructions.

```
        NDIAGS = 2 * N - 3
        ISTART = 1
        JSTART = 2
        LDIAG = 0
        DO 43141 IDIAGS = 1, NDIAGS
          IF (IDIAGS .LE. N - 1) THEN
            ISTART = ISTART + 1
            LDIAG = LDIAG + 1
          ELSE
            JSTART = JSTART + 1
            LDIAG = LDIAG - 1
          ENDIF
          I = ISTART + 1
          J = JSTART - 1
CDIR$ IVDEP
          DO 43142 IPOINT = 1, LDIAG
            I = I - 1
            J = J + 1
            A(I,J,1) = A(I,J,1) - B(I,J) * A(I-1,J,1)
    *                          - C(I,J) * A(I,J-1,1)
            A(I,J,2) = A(I,J,2) - B(I,J) * A(I-1,J,2)
```

```
  *                               - C(I,J) * A(I,J-1,2)
            A(I,J,3)=A(I,J,3)  - B(I,J) * A(I-1,J,3)
  *                               - C(I,J) * A(I,J-1,3)
43142 CONTINUE
43141 CONTINUE
```

The following chart shows the improvement on an older vector system from Cray versus the performance attained on the multicore architecture (Figure 6.5).

Note that the multicore using the SSE3 instructions never attains an improvement. Actually, the loop is not vectorized due to the striding through memory. However, the performance on the legacy vector system is quite significant. This type of restructuring, introducing a stride and/or indirect addressing will again be beneficial when applications are optimized for a general purpose graphics processor unit (GPGPU).

6.2.2 Data Dependence

To automatically vectorize a Fortran DO loop, the compiler must analyze the loop to ensure that the iterations are independent. The previous DO loop was an example of a DO loop that is really recursive. While there are algorithms that are inherently data dependent and cannot be vectorized

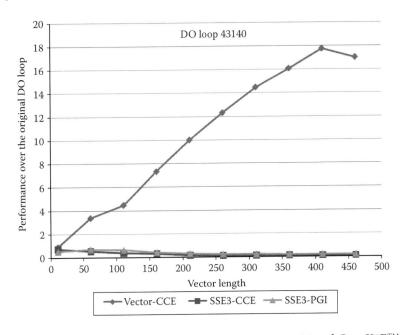

FIGURE 6.5   Comparison of DO 43140 restructuring on x86 and Cray X1E™.

without some restructuring, often the compiler may be fooled by the coding of a DO loop to believe that it has loop-carried dependencies when it does not. For example, consider the following:

```
C GAUSSIAN ELIMINATION
        DO 43020 I = 1, MATDIM
          A(I,I) = 1./A(I,I)
          DO 43021 J = I + 1, MATDIM
            A(J,I)=A(J,I) * A(I,I)
    43021 CONTINUE
          DO 43020 K = I + 1, MATDIM
            DO 43020 J = I + 1, MATDIM
              A(J,K) = A(J,K) - A(J,I) * A(I,K)
    43020 CONTINUE
```

In this loop nest, the innermost DO loop on J is data dependent if J and I can overlap, or if K and I can overlap. But the range of both J and K is I + 1 to MATDIM. So it is evident from our analysis that neither J nor K overlap with I within that range. Often when the programmer knows that the loop is not recursive and the compiler is confused, placing the !DIR$ IVDEP directive before the DO loop tells the compiler this loop is vectorizeable. While this is an old Cray traditional vector directive, most compilers recognize this to mean that it is safe to ignore apparent data dependencies.

The following DO loop is another example where the compiler cannot determine if the operations are dependent or not. If the array IA() has no repeated values, the loop operations are independent.

```
C     THE ORIGINAL
      DO 43070 I = 1, N
        A(IA(I)) = A(IA(I)) + C0 * B(I)
43070 CONTINUE
```

Now if we place the CDIR$ IVDEP in front of this loop, the compilers on the multicore systems will not vectorize this due to the indirect address. Once again this is an example where the legacy vector system benefits more from this restructuring. The legacy vector system gives us an idea of the benefit that will be achieved on future GPGPU systems that benefit more from vectorization of the loop (Figure 6.6).

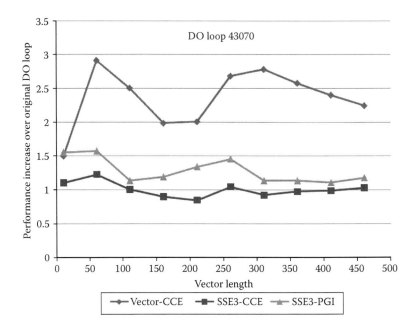

FIGURE 6.6   Comparison of DO 43070 restructuring on x86 and Cray X1E.

The next example is another DO loop that contains a recursive operation. First the Fortran:

```
C      THE ORIGINAL
       DO 43090 I = 2, N
       RLD = C(I)  -  B(I)
       RLDI = 1. / (RLD + .01)
       RLD1 = RLDI + 1.0
       D(I,1) = (D(I,1)  -  RLD1 * D(I,4))  * RLDI
       D(I,2) = (D(I,2)  -  RLD1 * D(I,4))  * RLDI
       D(I,3) = (D(I,3)  -  RLD1 * D(I,4))  * RLDI
       B(I) = (D(I,4)  -  RLD1 * A(I-1))  * RLDI
       A(I) = E(I)  * RLDI * B(I)
43090 CONTINUE
```

And now the C:

```
//     THE ORIGINAL
       for(I = 0;   I < N;   I++){
         RLD = C[I]  -  B[I];
         RLDI = 1. / RLD;
         RLD1 = RLDI + 1.0;
         D[0][I] = (D[0][I]  -  RLD1 * D[3][I])  * RLDI;
```

```
        D[1][I] = (D[1][I] - RLD1 * D[3][I]) * RLDI;
        D[2][I] = (D[2][I] - RLD1 * D[3][I]) * RLDI;
        B[I] = (D[3][I] - RLD1 * A[I-1]) * RLDI;
        A[I] = E[I] * RLDI * B[I];
    }
```

Note that A(I) is calculated from B(I) which in turn is dependent on A(I − 1). In this loop, the values of the A array that are computed in the Kth pass through the DO loop is used in the K + 1th pass through the DO loop. Fortunately, a majority of the computation is not recursive. It would be possible to vectorize a portion of this DO loop as follows:

In Fortran:

```
C       THE RESTRUCTURED
        DO 43091 I = 2, N
        RLD = C(I) - B(I)
        VRLDI(I) = 1./(RLD+.01)
        RLD1 = VRLDI(I) + 1.0
        D(I,1) = (D(I,1) - RLD1 * D(I,4)) * VRLDI(I)
        D(I,2) = (D(I,2) - RLD1 * D(I,4)) * VRLDI(I)
        D(I,3) = (D(I,3) - RLD1 * D(I,4)) * VRLDI(I)
43091 CONTINUE
        DO 43092 I = 2, N
          RLD1 = VRLDI(I) + 1.0
          B(I) = (D(I,4) - RLD1 * A(I-1)) * VRLDI(I)
          A(I) = E(I) * VRLDI(I) * B(I)
43092 CONTINUE
```

And in C:

```
//      THE RESTRUCTURED
        for(I = 0; I < N; I++) {
        RLD = C[I] - B[I];
        VRLDI[I] = 1./RLD;
        RLD1 = VRLDI[I] + 1.0;
        D[0][I] = (D[0][I] - RLD1 * D[3][I]) * VRLDI[I];
        D[1][I] = (D[1][I] - RLD1 * D[3][I]) * VRLDI[I];
        D[2][I] = (D[2][I] - RLD1 * D[3][I]) * VRLDI[I];
        }
        for(I = 0; I < N; I++) {
          RLD1 = VRLDI[I] + 1.0;
          B[I] = (D[3][I] - RLD1 * A[I-1]) * VRLDI[I];
          A[I] = E[I] * VRLDI[I] * B[I];
        }
```

The computations in the first DO loop are vectorizable and they have been split into a separate DO loop. While we achieve some vectorization in this case, overhead is introduced in the form of two looping constructs instead of one and we have caused some memory traffic due to the requirement to promote the RLD1 array to VRLD1 to carry information from the first to the second DO loop. In Figure 6.7, we see that the rewrite only achieved a small performance gain. The overhead we introduced was larger than the benefit from the vectorization.

There is another form of data recursion that is vectorizable and is used quite heavily in real scientific applications. A DOT_PRODUCT is a recursion operation that takes two vectors and reduces them down to a single scalar. The following DO loop is vectorizable:

```
      SUM = 0.0
      DO 44011 I = 1, N
      SUM = SUM + B(I) * C(I)
44011 CONTINUE
```

While the multiply in the DO loop can be performed with a full vector operation, the add is a pseudovector operation and does not achieve as

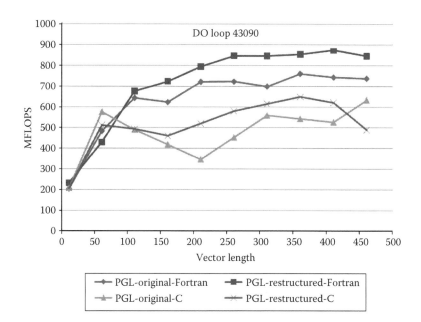

FIGURE 6.7 Comparison of original and restructured DO loop 43090 in Fortran and C.

much of a performance gain as a full vector operation. The analysis of a matrix multiply illustrates the difference between the performance of a DOT_PRODUCT and a full TRIAD vector operation (V = V + Scalar ⋆ V)

```
C      THE ORIGINAL
       DO 44050 I = 1, N
         DO 44050 J = 1, N
           A(I,J) = 0.0
           DO 44050 K = 1, N
             A(I,J)=A(I,J)+B(I,K)*C(K,J)
44050 CONTINUE
```

Consider the innermost K loop. This operation is a DOT_PRODUCT. A(I,J) is a scalar with respect to K. So we have an operation similar to the SUM loop above. There is also a slight complication that the B(I,K) array is accessed with a stride on K. An improved rewrite is

```
C      THE RESTRUCTURED
       DO 44051 J = 1, N
         DO 44051 I = 1, N
           A(I,J) = 0.0
44051 CONTINUE
       DO 44052 K = 1, N
         DO 44052 J = 1, N
           DO 44052 I = 1, N
             A(I,J)=A(I,J)+B(I,K)*C(K,J)
44052 CONTINUE
```

In Figure 6.8, the performance in MFLOPS is plotted against the length of the major loop index.

The loop contains a Vector (A(I,J)) = Vector(A(I,J)) + Vector(B(I,K))⋆ Scalar(C(K,J)) and the performance is vastly improved. Part of the performance gain is due to eliminating the stride in the original version of the loop. From looking at the output of the compile, the stride did not inhibit the compilers from vectorizing the original version of the DO loop.

The next example illustrates the use of a scalar that introduces recursion into the DO loop. This is typically referred to as a wrap around scalar. The value the scalar had in one pass of the DO loop is passed onto the next iteration.

```
       PF = 0.0
       DO 44030 I = 2, N
```

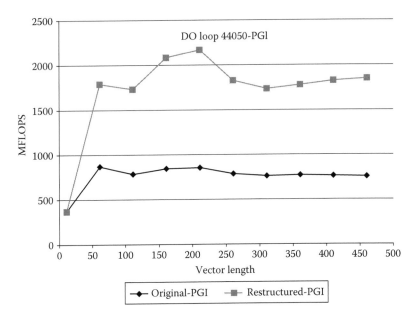

FIGURE 6.8   Comparison of matrix multiply rewrite.

```
        AV = B(I) * RV
        PB = PF
        PF = C(I)
        IF ((D(I) + D(I+1)) .LT. 0.)  PF = -C(I+1)
        AA = E(I) - E(I-1) + F(I) - F(I-1)
  1         + G(I) + G(I-1) - H(I) - H(I-1)
        BB = R(I) + S(I-1) + T(I) + T(I-1)
  1         - U(I) - U(I-1) + V(I) + V(I-1)
  2         - W(I) + W(I-1) - X(I) + X(I-1)
        A(I) = AV * (AA + BB + PF - PB + Y(I) - Z(I)) + A(I)
44030 CONTINUE
```

So, we have recursion and none of the compilers vectorize this DO loop. Now consider the following rewrite.

```
        VPF(1) = 0.0
        DO 44031 I = 2, N
        AV = B(I) * RV
        VPF(I) = C(I)
        IF ((D(I) + D(I+1)) .LT. 0.)  VPF(I) = -C(I+1)
        AA = E(I) - E(I-1) + F(I) - F(I-1)
  1         + G(I) + G(I-1) - H(I) - H(I-1)
```

```
       BB = R(I) + S(I-1) + T(I) + T(I-1)
  1        - U(I) - U(I-1) + V(I) + V(I-1)
  2          -W(I) + W(I-1) - X(I) + X(I-1)
       A(I) = AV * (AA + BB + VPF(I) - VPF(I - 1) + Y(I) - Z(I)) + A(I)
44031 CONTINUE
```

This rewrite is not recursive, since all the elements of the VPF are computed prior to their use in setting A(I). Once again we give the performance for the current multicores and the legacy vector system. The legacy system benefits much more from this restructuring (Figure 6.9).

6.2.3 IF Statements

Currently, none of the compilers vectorize DO loops containing IF statements. Once again, not because they cannot, but because the overhead of vectorizing the IF statement would make the resultant vector code slower than the original scalar code. Refer to the discussion in Chapter 3 with respect to the vectorization of IFs.

There are several types of IF statements that can easily be rewritten to vectorize with significant performance improvement. Consider the following lengthy example that illustrates how the programmer can optimize code containing IF statements.

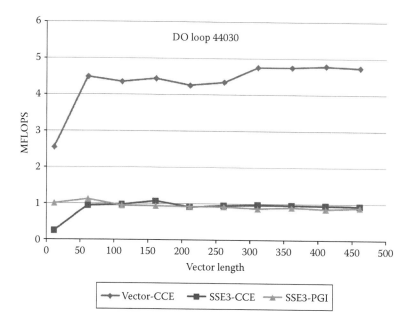

FIGURE 6.9   Comparison of DO 44030 restructuring on x86 and Cray X1E.

```
! The three mesh loops from the minimum to the maximum of each
  dimension
   DO 47020 J = 1, JMAX
     DO 47020 K = 1, KMAX
       DO 47020 I = 1, IMAX
        JP = J + 1
        JR = J - 1
        KP = K + 1
        KR = K - 1
        IP = I + 1
        IR = I - 1
! Test on J boundaries
        IF(J .EQ. 1) GO TO 50
        IF(J .EQ. JMAX) GO TO 51
!Only perform computation if J is an interior point
        XJ = (A(I,JP,K) - A(I,JR,K)) * DA2
        YJ = (B(I,JP,K) - B(I,JR,K)) * DA2
        ZJ = (C(I,JP,K) - C(I,JR,K)) * DA2
        GO TO 70
!Update for J = 1
  50 J1 = J + 1
     J2 = J + 2
     XJ = (-3. * A(I,J,K) + 4. * A(I,J1,K) - A(I,J2,K)) * DA2
     YJ = (-3. * B(I,J,K) + 4. * B(I,J1,K) - B(I,J2,K)) * DA2
     ZJ = (-3. * C(I,J,K) + 4. * C(I,J1,K) - C(I,J2,K)) * DA2
     GO TO 70
!Update for J = JMAX
  51 J1 = J - 1
     J2 = J - 2
     XJ = (3. * A(I,J,K) - 4. * A(I,J1,K) + A(I,J2,K)) * DA2
     YJ = (3. * B(I,J,K) - 4. * B(I,J1,K) + B(I,J2,K)) * DA2
     ZJ = (3. * C(I,J,K) - 4. * C(I,J1,K) + C(I,J2,K)) * DA2
70     CONTINUE
!Test on K boundaries
        IF(K .EQ. 1) GO TO 52
          IF(K .EQ. KMAX) GO TO 53
!Only perform computation if K is an interior point
        XK = (A(I,J,KP) - A(I,J,KR)) * DB2
        YK = (B(I,J,KP) - B(I,J,KR)) * DB2
        ZK = (C(I,J,KP) - C(I,J,KR)) * DB2
        GO TO 71
```

```
!Update for K = 1
   52 K1 = K + 1
      K2 = K + 2
      XK = (-3. * A(I,J,K) + 4. * A(I,J,K1) - A(I,J,K2)) * DB2
      YK = (-3. * B(I,J,K) + 4. * B(I,J,K1) - B(I,J,K2)) * DB2
      ZK = (-3. * C(I,J,K) + 4. * C(I,J,K1) - C(I,J,K2)) * DB2
      GO TO 71
!Update for K = KMAX
   53 K1 = K - 1
      K2 = K - 2
      XK = (3. * A(I,J,K) - 4. * A(I,J,K1) + A(I,J,K2)) * DB2
      YK = (3. * B(I,J,K) - 4. * B(I,J,K1) + B(I,J,K2)) * DB2
      ZK = (3. * C(I,J,K) - 4. * C(I,J,K1) + C(I,J,K2)) * DB2
71    CONTINUE
!Test on I boundaries
      IF(I .EQ. 1) GO TO 54
      IF(I .EQ. IMAX) GO TO 55
!Only perform computation if I is an interior point
      XI = (A(IP,J,K) - A(IR,J,K)) * DC2
      YI = (B(IP,J,K) - B(IR,J,K)) * DC2
      ZI = (C(IP,J,K) - C(IR,J,K)) * DC2
      GO TO 60
!Update for I = 1
   54 I1 = I + 1
      I2 = I + 2
      XI = (-3. * A(I,J,K) + 4. * A(I1,J,K) - A(I2,J,K)) * DC2
      YI = (-3. * B(I,J,K) + 4. * B(I1,J,K) - B(I2,J,K)) * DC2
      ZI = (-3. * C(I,J,K) + 4. * C(I1,J,K) - C(I2,J,K)) * DC2
      GO TO 60
!Update for I = IMAX
   55 I1 = I - 1
      I2 = I - 2
      XI = (3. * A(I,J,K) - 4. * A(I1,J,K) + A(I2,J,K)) * DC2
      YI = (3. * B(I,J,K) - 4. * B(I1,J,K) + B(I2,J,K)) * DC2
      ZI = (3. * C(I,J,K) - 4. * C(I1,J,K) + C(I2,J,K)) * DC2
60    CONTINUE
!Major computation
      DINV = XJ * YK * ZI + YJ * ZK * XI + ZJ * XK * YI
*           - XJ * ZK * YI - YJ * XK * ZI - ZJ * YK * XI
      D(I,J,K) = 1./(DINV + 1.E-20)
47020 CONTINUE
```

The end result of the loop is the calculation of D(I,J,K); however, there are numerous boundary conditions that keep the compilers from vectorizing this piece of code. The rewrite is quite simple. The individual boundaries are done separately within the K,J loop as follows:

Here are the J boundaries:

```
        DO 47029 J = 1, JMAX
          DO 47029 K = 1, KMAX
            IF(J.EQ.1)THEN
            J1 = 2
            J2 = 3
            DO 47021 I = 1, IMAX
            VAJ(I) = (-3. * A(I,J,K) + 4. * A(I,J1,K) - A(I,J2,K)) * DA2
            VBJ(I) = (-3. * B(I,J,K) + 4. * B(I,J1,K) - B(I,J2,K)) * DA2
            VCJ(I) = (-3. * C(I,J,K) + 4. * C(I,J1,K) - C(I,J2,K)) * DA2
47021 CONTINUE
            ELSE IF(J.NE.JMAX) THEN
            JP = J + 1
            JR = J - 1
            DO 47022 I = 1, IMAX
              VAJ(I) = (A(I,JP,K) - A(I,JR,K)) * DA2
              VBJ(I) = (B(I,JP,K) - B(I,JR,K)) * DA2
              VCJ(I) = (C(I,JP,K) - C(I,JR,K)) * DA2
47022 CONTINUE
            ELSE
            J1 = JMAX - 1
            J2 = JMAX - 2
            DO 47023 I = 1, IMAX
              VAJ(I) = (3. * A(I,J,K) - 4. * A(I,J1,K) + A(I,J2,K)) * DA2
              VBJ(I) = (3. * B(I,J,K) - 4. * B(I,J1,K) + B(I,J2,K)) * DA2
              VCJ(I) = (3. * C(I,J,K) - 4. * C(I,J1,K) + C(I,J2,K)) * DA2
47023 CONTINUE
            ENDIF
```

Note that temporary arrays (VAJ, VBJ, and VCJ) are introduced to carry the boundary values to the loop that calculates D(I,J,K). Here are the K boundaries:

```
        IF(K.EQ.1) THEN
        K1 = 2
        K2 = 3
```

```
      DO 47024 I = 1, IMAX
        VAK(I) = (-3. * A(I,J,K) + 4. * A(I,J,K1) - A(I,J,K2)) * DB2
        VBK(I) = (-3. * B(I,J,K) + 4. * B(I,J,K1) - B(I,J,K2)) * DB2
        VCK(I) = (-3. * C(I,J,K) + 4. * C(I,J,K1) - C(I,J,K2)) * DB2
47024 CONTINUE
      ELSE IF(K.NE.KMAX)THEN
      KP = K + 1
      KR = K - 1
      DO 47025 I = 1, IMAX
        VAK(I) = (A(I,J,KP) - A(I,J,KR)) * DB2
        VBK(I) = (B(I,J,KP) - B(I,J,KR)) * DB2
        VCK(I) = (C(I,J,KP) - C(I,J,KR)) * DB2
47025 CONTINUE
      ELSE
      K1 = KMAX - 1
      K2 = KMAX - 2
      DO 47026 I = 1, IMAX
        VAK(I) = (3. * A(I,J,K) - 4. * A(I,J,K1) + A(I,J,K2)) * DB2
        VBK(I) = (3. * B(I,J,K) - 4. * B(I,J,K1) + B(I,J,K2)) * DB2
        VCK(I) = (3. * C(I,J,K) - 4. * C(I,J,K1) + C(I,J,K2)) * DB2
47026 CONTINUE
      ENDIF
```

And then, when we have a boundary condition on I, we should simply leave them outside the I loop as follows:

```
      I = 1
      I1 = 2
      I2 = 3
      VAI(I) = (-3. * A(I,J,K) + 4. * A(I1,J,K) - A(I2,J,K)) * DC2
      VBI(I) = (-3. * B(I,J,K) + 4. * B(I1,J,K) - B(I2,J,K)) * DC2
      VCI(I) = (-3. * C(I,J,K) + 4. * C(I1,J,K) - C(I2,J,K)) * DC2
      DO 47027 I = 2, IMAX-1
        IP = I + 1
        IR = I - 1
        VAI(I) = (A(IP,J,K) - A(IR,J,K)) * DC2
        VBI(I) = (B(IP,J,K) - B(IR,J,K)) * DC2
        VCI(I) = (C(IP,J,K) - C(IR,J,K)) * DC2
47027 CONTINUE
      I = IMAX
      I1 = IMAX - 1
      I2 = IMAX - 2
```

```
      VAI(I) = (3. * A(I,J,K) - 4. * A(I1,J,K) +A(I2,J,K)) * DC2
      VBI(I) = (3. * B(I,J,K) - 4. * B(I1,J,K) +B(I2,J,K)) * DC2
      VCI(I) = (3. * C(I,J,K) - 4. * C(I1,J,K) +C(I2,J,K)) * DC2
```

And finally, the end of the K,J loop:

```
      DO 47028 I = 1, IMAX
      DINV = VAJ(I) * VBK(I) * VCI(I) +VBJ(I) * VCK(I) * VAI(I)
   1       +VCJ(I) * VAK(I) * VBI(I) - VAJ(I) * VCK(I) * VBI(I)
   2       - VBJ(I) * VAK(I) * VCI(I) - VCJ(I) * VBK(I) * VAI(I)
      D(I,J,K) = 1. / (DINV + 1.E-20)
47028 CONTINUE
47029 CONTINUE
```

Now every DO loop is vectorized. By splitting the loop, we end up increasing the amount of memory movement. To minimize this, make sure to only split out of the innermost loop to reduce the amount of data that will be used from loop to loop. We have introduced numerous temporary arrays; however, the temporaries are only dimensioned on I, which means that these small arrays will probably stay in cache and not increase memory bandwidth. Given this optimization, we obtain a very significant performance gain, as shown in Figure 6.10.

The following example shows another loop with IF statements which can be optimized with a little rewrite:

```
C   THE ORIGINAL
    DO 47030 I = 1, N
        A(I) = PROD * B(1,I) * A(I)
        IF (A(I) .LT. 0.0) A(I) = -A(I)
        IF (XL .LT. 0.0) A(I) = -A(I)
        IF (GAMMA) 47030, 47030, 100
100     XL = -XL
47030     CONTINUE
```

This is from a real application. It also contains an ancient form of IF. The IF(GAMMA) statement goes to 47930 if GAMMA is less than zero or zero, and 100 if it is positive. The rewrite eliminates all the IFs.

```
    DO 47031 I = 1, N
       A(I) = PROD * B(1,I) * A(I)
       A(I) = ABS (A(I))
47031 CONTINUE
```

```
IF (GAMMA .LE. 0.) THEN
    IF (XL .LT. 0.0) THEN
       DO 47032 I = 1, N
          A(I) = -A(I)
47032 CONTINUE
    ENDIF
  ELSE
  IF (XL .LT. 0.0) THEN
     DO 47033 I = 1, N, 2
       A(I) = -A(I)
47033 CONTINUE
  ENDIF
  IF (XL .GT. 0.0) THEN
     DO 47034 I = 2, N, 2
        A(I) = -A(I)
47034 CONTINUE
    ENDIF
  ENDIF
```

Note the use of the ABS function to eliminate an IF in DO loop 47031. Other functions that can be used to eliminate IFs are MIN and MAX.

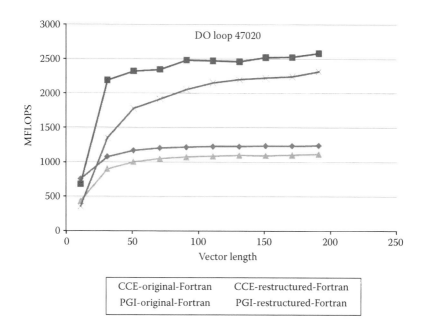

FIGURE 6.10   Comparison of original and restructured DO loop 47020 on CCE and PGI compilers.

These functions are highly optimized by the compiler and are good to use. The other IFs are loop-independent IFs, which should always be split out of DO loops (Figure 6.11).

An example of a very complicated IF is a traditional table look-up with interpolation. Given the previous discussion concerning the vectorization of indirect addressing, the restructuring will not perform as good as it once did on a legacy vector machine.

```
      DO 47101 I = 1, N
        U1 = X2(I)
        DO 47100 LT = 1, NTAB
          IF (U1 .GT. X1(LT)) GO TO 47100
            IL = LT
            GO TO 121
47100 CONTINUE
        IL = NTAB - 1
121   Y2(I) = Y1(IL) + ( Y1(IL + 1) - Y1(IL))/
      *                 (X1(IL + 1) - X1(IL)) *
      *                 (X2(I) - X1(IL))
47101 CONTINUE
```

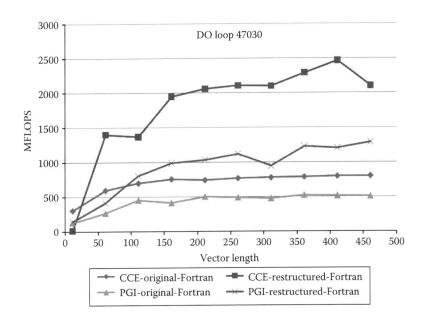

FIGURE 6.11 Comparison of original and restructured versions of DO loop 47030.

The major loop is the I loop that might be looping over the grid blocks in an application. The loop on LT is over the table, finding the interval in the table that contains the quantity X2(I). In the restructured loop, the look-up is separated from the actual interpolation.

```
      DO 47103 I = 1, N
        U1 = X2 (I)
        DO 47102 LT = 1, NTAB
          IF (U1 .GT. X1(LT)) GO TO 47102
            IV(I) = LT
            GO TO 47103
47102 CONTINUE
          IV(I) = NTAB - 1
47103 CONTINUE
      DO 47104 I = 1, N
        Y2(I) = Y1(IV(I)) + ( Y1(IV(I) + 1) - Y1(IV(I))) /
     *                  (X1(IV(I) + 1) - X1(IV(I))) *
     *                  (X2(I) - X1(IV(I)))
47104 CONTINUE
```

The look-up is performed and the values of the interval are saved into a temporary array IV. This is the overhead. The creation of temporary arrays force more memory motion and thus, the improvement obtained in the restructuring will suffer. The DO 47104 loop then performs the interpolation to find the desired Y2. This particular example was a clear win in the days of the old powerful vector processors. Given the SSE instruction performance, this restructuring might not be recommended for current SSE vector instructions (Figure 6.12).

The following IF construct is difficult to vectorize, because there is not even a DO loop. This is a good example of a computation that is performed until a certain criterion is reached. In this example, we perform a rewrite that performs more operations than are necessary and then only save the ones we require.

```
! ORIGINAL
    I = 0
47120 CONTINUE
    I = I + 1
    A(I) = B(I)**2 + .5 * C(I) * D(I) / E(I)
    IF (A(I) .GT. 0.) GO TO 47120
```

```
! RESTRUCTURED
  DO 47123 II = 1, N, 128
    LENGTH = MIN0 (128, N-II+1)
    DO 47121 I = 1, LENGTH
      VA(I) = B(I+II-1) ** 2 + .5 * C(I+II-1) * D(I+II-1) /
                                   E(I+II-1)
47121 CONTINUE
    DO 47122 I = 1, LENGTH
      A(I+II-1) = VA(I)
      IF (A(I+II-1) .LE. 0.0) GO TO 47124
47122 CONTINUE
47123 CONTINUE
47124 CONTINUE
```

DO loop 47121 vectorizes. However, if the first value of A less than or equal to zero appears early in the 47122 loop, we wind up doing a lot of unnecessary computations in the 47123 loop. Fewer wasted computations are done if the value of A that is less than or equal to zero occurs later in the loop.

The outer DO 47123 DO loop is called a strip mining loop. The computation is divided into smaller chunks, in this case 128, to minimize the

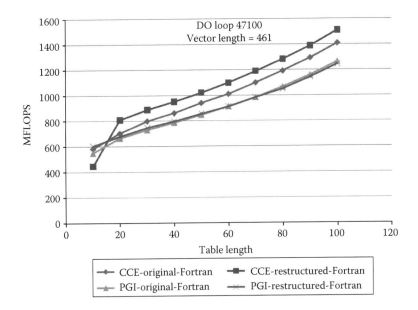

FIGURE 6.12   Comparison of original and restructured DO loop 47100.

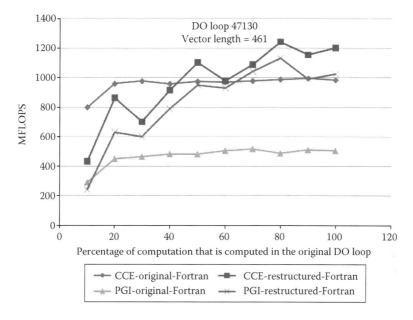

FIGURE 6.13   Comparison of the original and restructured DO 47130.

amount of unnecessary operations. However, strip mining can also be used to increase cache reuse (to be discussed later). As we can see from Figure 6.13, with the Cray compiler, the restructured does outperform the original when at least 50% of the loop is updated, and with the PGI compiler, the restructuring wins whenever some of the computation is performed.

What about the legacy vector system? Figure 6.14 illustrates that the performance gain achieved on the vector system is significantly better than that achieved on the systems with the SSE3 instructions.

6.2.4 Subroutine and Function Calls

Subroutine and function calls from within a DO inhibit vectorization. Many cases where the called routines are inlined still fail to vectorize. To overcome the principal problem, programmers should think of calling a routine to update many elements of an array. In this way, the loop that should be vectorized is contained within the routine of interest. When an application consists of multinested DO loops, only one of the loops needs to be pulled down into the computational routines. Ideally, this loop will update at least 100–200 elements of the array. Calling a subroutine to perform an update on a single grid point is very inefficient. We also see later how higher level loops are good for OpenMP.

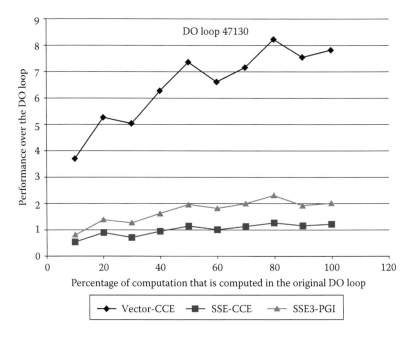

FIGURE 6.14   Comparisons of DO 47130 on x86 and Cray X1E.

To take an existing application and rewrite important DO loops that call subroutines, there are a number of options. First if the function/routine can be written as a single line update, consider using a statement function. Statement functions maintain modularity and the compiler expands the statement function within the DO loop and then vectorizes the DO loop. Consider the following example:

```
      DO 48020 I = 1, N
         A(I) = B(I) * FUNC(D(I)) + C(I)
48020 CONTINUE

      FUNCTION FUNC (X)
      FUNC = X**2 + 2.0 / X
      RETURN
      END
```

This function call can easily be turned into the following statement function:

```
C
      FUNCX(X) = X**2 + 2.0/X
```

```
      DO 48021 I = 1, N
        A(I) = B(I) * FUNCX (D(I)) + C(I)
48021 CONTINUE
```

And the compiler generates excellent vector code (Figure 6.15). Consider the following example:

```
      DO 48060 I = 1, N
        AOLD = A(I)
        A(I) = UFUN (AOLD, B(I), SCA)
        C(I) = (A(I) + AOLD) * .5
48060 CONTINUE
```

Which is rewritten:

```
      DO 48061 I = 1, N
        VAOLD(I) = A(I)
48061 CONTINUE
      CALL VUFUN (N, VAOLD, B, SCA, A)
      DO 48062 I = 1, N
        C(I) = (A(I) + VAOLD(I)) * .5
48062 CONTINUE
```

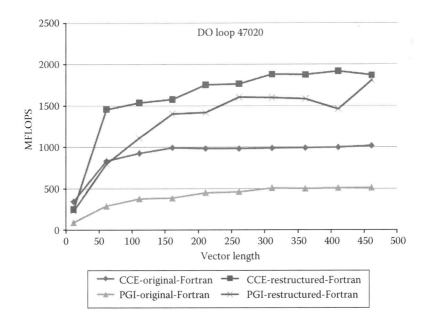

FIGURE 6.15 Comparisons of original and restructured DO 47020.

Note that this adds the overhead of three separate loops, two in the calling routine and one in the called routine, and a temporary one is used to pass information to the called routine.

The routine itself is

```
      FUNCTION UFUN (X, Y, SCA)
      IF (SCA .GT. 1.0) GO TO 10
        UFUN = SQRT (X**2 + Y**2)
        GO TO 5
10    UFUN = 0.0
5     CONTINUE
      RETURN
      END
```

Which is rewritten to

```
      SUBROUTINE VUFUN (N, X, Y, SCA, UFUN)
      DIMENSION X(*), Y(*), UFUN(*)
      IF (SCA .GT. 1.0) GO TO 15
      DO 10 I = 1, N
         UFUN(I) = SQRT (X(I)**2 + Y(I)**2)
10    CONTINUE
      RETURN
15    CONTINUE
      DO 20 I = 1, N
         UFUN(I) = 0.0
20    CONTINUE
      RETURN
      END
```

In this case, we obtain an added benefit that the IF becomes independent of the loop and can be taken outside the loop. Consequently, the optimization gives a good performance gain (Figure 6.16).

Consider the following example:

```
      COMMON /SCALAR/SCALR
      DO 48080 I = 1, N
      A(I) = SQRT(B(I)**2 + C(I)**2)
      SCA = A(I)**2 + B(I)**2
      SCALR = SCA*2
      CALL SUB1(A(I),B(I),SCA)
      CALL SUB2(SCA)
```

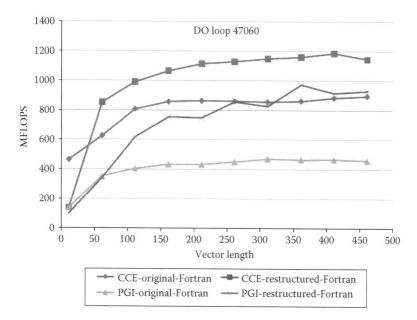

FIGURE 6.16   Comparisons of original and restructured DO 47060.

```
        D(I) = SQRT(ABS(A(I) + SCA))
48080 CONTINUE

        SUBROUTINE SUB1 (X, Y, SCA)
        COMMON /SCALAR/ SCALR
        SCA = X**2 + Y**2
        SCALR = SCA * 2
        RETURN
        END
        SUBROUTINE SUB2 (SCA)
        COMMON /SCALAR/ SCALR
        SCA = SCA + SCALR
        RETURN
        END
```

Here we have a more complicated example with two calls and the communication of the scalar variable through a COMMON block. The restructured version is:

```
        DO 48081 I = 1, N
          A(I) = SQRT(B(I)**2 + C(I)**2)
48081 CONTINUE
        CALL VSUB1(N,A,B,VSCA,VSCALR)
        CALL VSUB2(N,VSCA,VSCALR)
```

```
      DO 48082 I = 1, N
      D(I) = SQRT(ABS(A(I) + VSCA(I)))
48082 CONTINUE
    SUBROUTINE VSUB1 (N, X, Y, SCA, VSCALR)
      DIMENSION X(*), Y(*), SCA(*), VSCALR(*)
      COMMON /SCALAR/ SCALR
    DO 10 I = 1, N
      SCA(I) = X(I)**2 +Y(I)**2
      VSCALR(I) = SCA(I) * 2
10 CONTINUE
    SCALR = VSCALR(N)
    RETURN
    END
    SUBROUTINE VSUB2 (N, SCA, VSCALR)
    DIMENSION SCA(*), VSCALR(*)
    COMMON /SCALAR/ SCALR
    DO 10 I = 1, N
      SCA(I) = SCA(I) + VSCALR(I)
10 CONTINUE
    RETURN
    END
```

In this restructuring, care must be taken to assure that any variable that carries information from one loop to another is promoted to an array. Also note that the last value of SCALR was saved after DO loop 10 in subroutine VSUB1. This is to take care of the case when this COMMON block is contained in another routine in the program. Restructuring enables better performance on both the compilers (Figure 6.17).

In summary, when an application calls a routine from within a DO loop, the programmer can either (1) turn it into a statement function, (2) manually inline the called routine(s), or (3) split the DO loop at the call(s) and call a vector version of the routine. The following table gives some hints on when to do which one.

	Statement Function	Inline	Vector Routine
Simple	X	X	
Very large routine calling other routines			X
Very small		X	
When complications exist with COMMON and/or MODULES			X

6.2.4.1 Calling Libraries

Many code developers go overboard on calling libraries such as the basic linear algebra (BLAS). These important libraries are optimized by the various vendors to perform well on their target system; however, that optimization is targeted at large vector lengths. Following is a very computationally intensive routine from the WUPWISE SPEC® benchmark suite whose OpenMP parallelization is discussed in Chapter 6. In the gammul routine in the WUPWISE application, the developer called numerous BLAS routines to perform very short (vector length) operations:

```
IF (MU.EQ.1) THEN
CALL ZCOPY(12, X, 1, RESULT, 1)
CALL ZAXPY(3, I, X(10), 1, RESULT(1), 1)
CALL ZAXPY(3, I, X( 7), 1, RESULT(4), 1)
CALL ZAXPY(3, -I, X( 4), 1, RESULT(7), 1)
CALL ZAXPY(3, -I, X( 1), 1, RESULT(10), 1)

ELSE IF (MU.EQ.2) THEN
CALL ZCOPY(12, X, 1, RESULT, 1)
CALL ZAXPY(3, ONE, X(10), 1, RESULT(1), 1)
CALL ZAXPY(3, -ONE, X(7), 1, RESULT(4), 1)
```

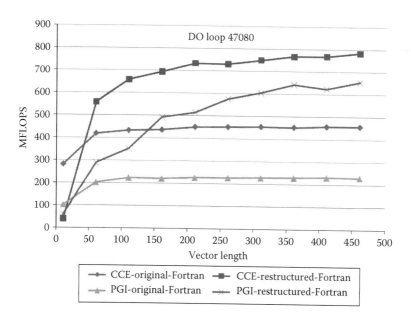

FIGURE 6.17   Comparisons of original and restructured DO 47080.

```
CALL ZAXPY(3, -ONE, X(4), 1, RESULT(7), 1)
CALL ZAXPY(3, ONE, X(1), 1, RESULT(10), 1)
o    o    o
```

There is a code much more similar to this which can be rewritten as

```
IF (MU.EQ.1) THEN
DO II = 1,12
   RESULT(II) = X(II)
ENDDO
DO II = 1,3
   RESULT(II) = RESULT(II) + I * X(II + 9)
   RESULT(II + 3) = RESULT(II + 3) + I * X(II + 6)
   RESULT(II + 6) = RESULT(II + 6) - I * X(II + 3)
   RESULT(II + 9) = RESULT(II + 9) - I * X(II)
ENDDO
ELSE IF (MU.EQ.2) THEN
DO II = 1,12
   RESULT(II) = X(II)
ENDDO
DO II = 1,3
   RESULT(II) = RESULT(II) + ONE * X(II + 9)
   RESULT(II + 3) = RESULT(II + 3) - ONE * X(II + 6)
   RESULT(II + 6) = RESULT(II + 6) - ONE * X(II + 3)
   RESULT(II + 9) = RESULT(II + 9) + ONE * X(II)
ENDDO
o    o    o
```

This particular optimization decreased the overall runtime for the WUPWISE benchmark on a single core from 459 to 409 s.

On the other hand, it is important to call the vendor supplied BLAS libraries for all significant computation using the BLAS routines. For example, the WUPWISE benchmark used its own Fortran version of the BLAS routines in the benchmark for calls that were much longer. This is probably for portability issues; however, when those Fortran versions are replaced with the vendor-supplied optimized routines, the benchmark runs significantly faster on a single core.

6.2.5 Multinested DO Loops

Multinested DO loops introduce a variety of optimization opportunities. Compilers tend to concentrate on the innermost DO loop, especially if there is a code between the inner and other DO loops. The most widely used multinested loop is probably the matrix multiply, a version of which is shown below.

```
C    THE ORIGINAL
     DO 46011 J = 1,  4
        DO 46010 I = 1,  N
           C(J,I) = 0.0
46010 CONTINUE
        DO 46011 K = 1,4
        DO 46011 I = 1,N
           C(J,I) = C(J,I) + A(J,K)  *  B(K,I)
46011 CONTINUE
```

In this case, one of the dimensions is 4. This loop performs a $(4 \times 4) \times (4 \times N)$ matrix multiply which results in a $4 \times N$ matrix. A general rule when having small DO loops around larger loops is to unroll the smaller loops inside the larger loops. In this case, there are two loops of length 4. The restructuring gives us the following:

```
DO 46012 I = 1,  N
C(1,I) = A(1,1)  *  B(1,I) + A(1,2)  *  B(2,I)
*          + A(1,3)  *  B(3,I) + A(1,4)  *  B(4,I)
C(2,I) = A(2,1)  *  B(1,I) + A(2,2)  *  B(2,I)
*          + A(2,3)  *  B(3,I) + A(2,4)  *  B(4,I)
C(3,I) = A(3,1)  *  B(1,I) + A(3,2)  *  B(2,I)
*          + A(3,3)  *  B(3,I) + A(3,4)  *  B(4,I)
C(4,I) = A(4,1)  *  B(1,I) + A(4,2)  *  B(2,I)
*          + A(4,3)  *  B(3,I) + A(4,4)  *  B(4,I)
46012 CONTINUE
```

The main problem with the restructuring is the stride in the input and result array. So this loop will not vectorize; however, the restructured will perform better than the original as shown in Figure 6.18.

Consider the next example which is very similar. However, the longest matrix dimension does not stride through the array.

In the Fortran version:

```
C    THE ORIGINAL
     DO  46020 I = 1,N
        DO  46020 J = 1,4
           A(I,J) = 0.
```

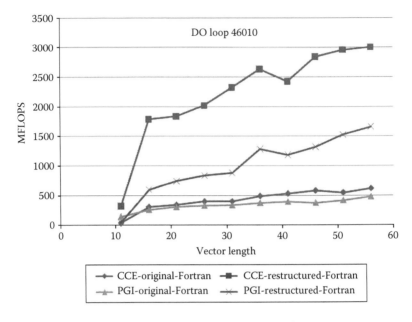

FIGURE 6.18   Comparisons of original and restructured DO 46010.

```
      DO 46020 K=1,4
         A(I,J) =A(I,J) +B(I,K)  *  C(K,J)
46020 CONTINUE
```

And now the C version:

```
//  THE ORIGINAL
    for(I = 0;   I < N;   I ++){
    for(J = 0;   J < 4;   J ++){
      A[J][I] = 0.;
      for(K = 0;   K < 4;   K ++){
         A[J][I] =A[J][I] +B[K][I]  *  C[J][K];
      }
    }
    }
```

Now when the unrolling of the short loops on four are performed, we obtain the following code:

The Fortran version:

```
C   THE RESTRUCTURED
    DO 46021 I = 1, N
       A(I,1) =B(I,1)  *  C(1,1) +B(I,2)  *  C(2,1)
    *           +B(I,3)  *  C(3,1) +B(I,4)  *  C(4,1)
```

```
     A(I,2) = B(I,1)  *  C(1,2) + B(I,2)  *  C(2,2)
    *            + B(I,3)  *  C(3,2) + B(I,4)  *  C(4,2)
     A(I,3) = B(I,1)  *  C(1,3) + B(I,2)  *  C(2,3)
    *            + B(I,3)  *  C(3,3) + B(I,4)  *  C(4,3)
     A(I,4) = B(I,1)  *  C(1,4) + B(I,2)  *  C(2,4)
    *            + B(I,3)  *  C(3,4) + B(I,4)  *  C(4,4)
46021 CONTINUE
```

And in C:

```
// THE RESTRUCTURED
   for (I = 0;   I < N;   I ++) {
     A[0] [I] = B[0] [I]  *  C[0] [0] + B[1] [I]  *  C[0] [1]
                  + B[2] [I]  *  C[0] [2] + B[3] [I]  *  C[0] [3];
     A[1] [I] = B[0] [I]  *  C[1] [0] + B[1] [I]  *  C[1] [1]
                  + B[2] [I]  *  C[1] [2] + B[3] [I]  *  C[1] [3];
     A[2] [I] = B[0] [I]  *  C[2] [0] + B[1] [I]  *  C[2] [1]
                  + B[2] [I]  *  C[2] [2] + B[3] [I]  *  C[2] [3];
     A[3] [I] = B[0] [I]  *  C[3] [0] + B[1] [I]  *  C[3] [1]
                  + B[2] [I]  *  C[3] [2] + B[3] [I]  *  C[3] [3];
   }
```

Our next example is the square matrix multiply. This was covered earlier in this section when discussing the elimination of the DOT_PRODUCT on the innermost DO loop. In the optimized libraries on the x86 systems, the library developers employ unrolling of loops to improve the matrix multiply even more. Following is the optimization from earlier in this chapter:

```
   DO 46030 J = 1,  N
      DO 46030 I = 1,  N
         A(I,J) = 0.
46030 CONTINUE
   DO 46031 K = 1,  N
      DO 46031 J = 1,  N
         DO 46031 I = 1,  N
            A(I,J) = A(I,J) + B(I,K)  *  C(K,J)
46031 CONTINUE
```

Here is the C code for this example:

```
//      THE ORIGINAL
   for (J = 0;   J < N;   J ++) {
   for (I = 0;   I < N;   I ++) {
```

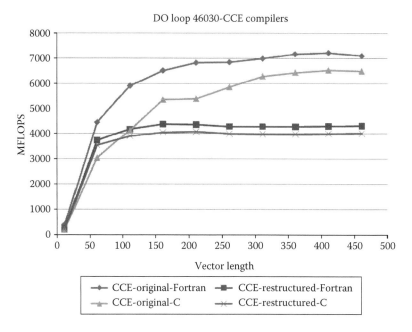

FIGURE 6.19   Comparisons of original and restructured DO 46030 in Fortran and C.

```
      A[J][I] = 0.0;
    }
  }
  for(K = 0;   K < N;   K++) {
  for(J = 0;   J < N;   J++) {
  for(I = 0;   I < N;   I++) {
    A[J][I] = A[J][I] + B[K][I] * C[J][K];
  }
  }
  }
```

When this is run through the two compilers, the performance obtained is not as good as could be expected. Typically, a double precision matrix multiply (DGEMM) for any good size matrix should obtain at least 50% of the peak. This example gives us some hints on how we can improve other multinested DO loops. In this example, the large majority of the time is spent in the single statement in the triple-nested 46031 loop. This statement does not have much computation intensity and is dominated by the two loads of A(I,J) and B(I,K). The computational intensity of this loop is 2 (add + multiply)/(2 loads + 1 store) = 2/3. How can this ratio that is directly proportional to the performance of the loop be increased? In the last example, we unrolled small loops inside the inner loop and increased the number of computations performed in the inner DO loop. While we do not have a small outer loop, there are two loops that can be done in chunks, the J and K loops. Consider the following restructuring:

In the Fortran version:

```
   DO 46032 J = 1, N
      DO 46032 I = 1, N
         A(I,J) = 0.
46032 CONTINUE
C
   DO 46033 K = 1, N-5, 6
      DO 46033 J = 1, N
         DO 46033 I = 1, N
            A(I,J) = A(I,J) + B(I,K)  *  C(K,J)
     *                      + B(I,K+1) * C(K+1,J)
     *                      + B(I,K+2) * C(K+2,J)
     *                      + B(I,K+3) * C(K+3,J)
     *                      + B(I,K+4) * C(K+4,J)
     *                      + B(I,K+5) * C(K+5,J)
46033 CONTINUE
C
   DO 46034 KK = K, N
      DO 46034 J = 1, N
         DO 46034 I = 1, N
            A(I,J) = A(I,J) + B(I,KK)  * C(KK,J)
46034 CONTINUE
```

And the C code:

```
//  THE RESTRUCTURED
   for(J = 0;   J < N;   J++) {
   for(I = 0;   I < N;   I++) {
      A[J][I] = 0.0;
   }
   }
   for(K = 0;   K < N-5;   K = K+6) {
   for(J = 0;   J < N;   J++) {
   for(I = 0;   I < N;   I++) {
      A[J][I] = A[J][I] + B[K][I]  *  C[J][K]
                        + B[K+1][I] * C[J][K+1]
```

```
                    + B [K + 2] [I]  *  C [J] [K + 2]
                    + B [K + 3] [I]  *  C [J] [K + 3]
                    + B [K + 4] [I]  *  C [J] [K + 4]
                    + B [K + 5] [I]  *  C [J] [K + 5];
   }
   }
   }
   for (KK = K;   KK < N;   KK ++) {
   for (J = 0;   J < N;   J ++) {
   for (I = 0;   I < N;   I ++) {
      A [J] [I] = A [J] [I] + B [KK] [I]  *  C [J] [KK];
   }
   }
   }
```

In this restructuring, the K loop is split up into chunks of six, and the six are unrolled inside the I loop. For each added K iteration, a single load and two flops are added. Therefore, the computational intensity of the restructured loop is 12/8 = 1.5. Looking at Figure 6.21, we see that we slowed down this example. The reason is that the compiler put in the call to DGEMM in

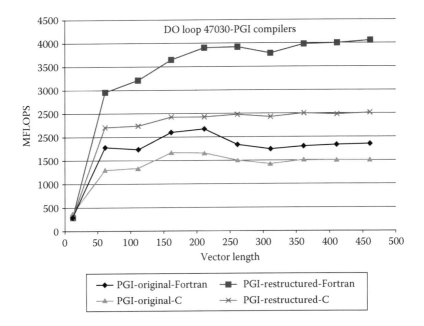

FIGURE 6.20 Comparisons of original and restructured DO 47030 in Fortran and C.

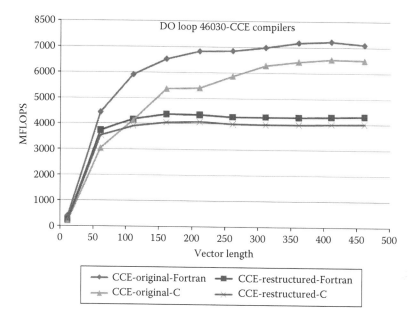

FIGURE 6.21   Comparisons of original and restructured DO 46030 in Fortran and C.

the original and our restructured has obscured the matrix multiply so much, that the compiler could not replace the restructured with a DGEMM call. Although this runs slower, it should not discourage the user from using this powerful unrolling technique on examples where the compiler does not perform the pattern match of the matrix multiply kernel.

Note that the last triple-nested DO loop performs the cleanup. If N is not a multiple of six, then there would be a remainder to compute. This cleanup loop does not use too much time; hence we do not have to be concerned about restructuring it.

How much unrolling should be used? As we unroll, more compiler temporaries are generated and saved in registers. When the number of registers are exhausted, there are spills to cache and that causes a degradation in performance. Experimenting with the unroll amount is the best way to determine what amount to use.

The restructuring shown in this example is exactly what the library developers employ to create a more efficient matrix multiply. In cases where the matrices are square and relatively large, it is best to call the library equivalent of matrix multiply. Whenever one of the dimensions of the matrices is small, say smaller than 40–50, it may be more efficient to hand optimize the Fortran as is being done here. Note in Figure 6.21 that the Cray compiler recognizes the original kernel as a matrix product, and the restructuring does not allow this optimization. The PGI compiler does not have the ability to recognize the matrix multiply in the restructuring (Figures 6.19 through 6.21).

The performance that the Cray compiler achieves in this example indicates that the compiler did something extra with the restructured DO loop. The following compiler-generated listing shows that the compiler replaced the matrix multiply with a call to an optimized library routine:

```
43. 1 2------< DO 46030 J = 1,  N

44. 1 2 A-----< DO 46030 I = 1,  N

45. 1 2 A A(I,J) = 0.

46. 1 2 A----- >> 46030 CONTINUE

47. 1

48. 1 A------< DO 46031 K = 1,  N

49. 1 A 3-----< DO 46031 J = 1,  N

50. 1 A 3 4---< DO 46031 I = 1,  N

51. 1 A 3 4 A(I,J) = A(I,J) + B(I,K)  *  C(K,J)

52. 1 A 3 4--- > >> 46031 CONTINUE
```

The listing indicates with the A that the looping structure was pattern matched and replaced with a call to DGEMM. More matrix multiply kernels will be examined later in this section.

When an application has multinested loop structures, the ideal situation is that the innermost loop accesses the data contiguously with stride one, small loops are unrolled, and the second largest loop is the outer loop and accesses the array on the outermost dimension. This structure gives the best inner loops for vectorization and the outermost DO loop can be optimized with OpenMP.

At times, the application may not have the combination mentioned above. Consider the following looping structure. If N is greater than 10, the largest loop is on the outside. Unfortunately, the loop that accesses C contiguously is not very long.

```
DO 45020 I = 1,  N
    F(I) = A(I) + .5
```

```
      DO 45020 J=1, 10
         D(I,J) = B(J) * F(I)
         DO 45020 K=1, 5
            C(K,I,J) = D(I,J) * E(K)
45020 CONTINUE
```

If we take the approach of pulling the longest loop on the inside, then we have a stride on the C array. Since the K index is short, and I is the second index it will still make good utilization of the cache. If the first dimension of C is 5, then 5*N contiguous elements of C are being used within the DO loop. Consider the restructuring that pulls the I loop inside the other loops, Since N is expected to be the largest of the loops, this would result in the longest loop on the inside.

```
      DO 45021 I=1,N
         F(I) = A(I) + .5
45021 CONTINUE
      DO 45022 J=1, 10
         DO 45022 I=1, N
            D(I,J) = B(J) * F(I)
45022 CONTINUE
      DO 45023 J=1, 10
         DO 45023 K=1, 5
         DO 45023 I=1, N
            C(K,I,J) = D(I,J) * E(1)
45023 CONTINUE
```

Figure 6.22 shows the improvement.

6.2.5.1 More Efficient Memory Utilization

6.2.5.1.1 Loop Tiling for Cache Reuse   The following differencing algorithm on a 2D or 3D grid illustrates a good example where cache blocking or loop tiling is beneficial. Here is a triple-nested DO loop from the RESID routine in the MGRID application in the NASA parallel benchmark suite (NPB):

```
DO I3=2, N3-1
   DO I2=2, N2-1
      DO I1=1, N1
         U1(I1) = U(I1,I2-1,I3) + U(I1,I2+1,I3)
>                 + U(I1,I2,I3-1) + U(I1,I2,I3+1)
```

```
            U2(I1) = U(I1,I2-1,I3-1) + U(I1,I2+1,I3-1)
>                  + U(I1,I2-1,I3+1) + U(I1,I2+1,I3+1)
       ENDDO
       DO I1 = 2,N1-1
          R(I1,I2,I3) = V(I1,I2,I3)
>            - A(0) * U(I1,I2,I3)
>            - A(2) * (U2(I1) + U1(I1-1) + U1(I1+1))
>            - A(3) * (U2(I1-1) + U2(I1+1))
       ENDDO
    ENDDO
ENDDO
```

If we consider the entire grid on which this algorithm executes, each interior point is updated using $I1 \pm 1$, $I2 \pm 1$, and $I3 \pm 1$. Figure 6.23 illustrates this on a 3D grid.

As the first plane is updated, operands are brought into cache as each new i1 line is required. During this process, the U values required in the $I2 + 1$ line and $I3 + 1$ plane are also brought into cache. Once we complete the first I3 plane, we will have most of the u array for the first plane in cache. As we work on the second plane, we will bring in the third I3 plane. As we can see, we should get cache reuse of a u line nine times, as (I2,I3),

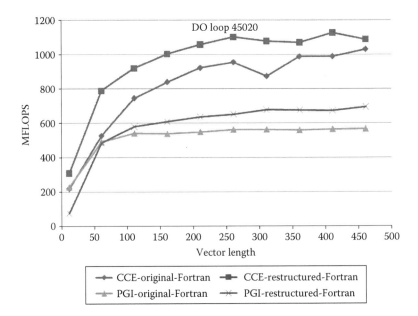

FIGURE 6.22   Comparisons of original and restructured DO 45020.

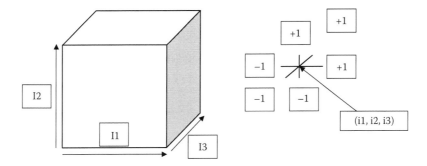

FIGURE 6.23 Visualization of 3D stencil in RESID routine from MGRID.

(I2 + 1,I3), (I2 − 1,I3), (I2,I3 + 1), (I2 + 1,I3 + 1), (I2 − 1,I3 + 1), (I2,I3 + 1), (I2 + 1,I3 + 1), and (I2 − 1,I3 + 1). This is excellent cache reuse, unless our cache is too small.

In the previous section, we saw that Level 1 cache would hold 64KB, or 8192 8-byte words. Since we are using three arrays in the DO loop we can calculate the subgrid size that would fit into Level 1 cache (8192/3 = 2730 or about 14^3). This size blocking is not good, because the inner DO loop would be updating less than two cache lines. Another approach would be to block for Level 2 cache. This cache is somewhat larger (512 KB on quad-core Opteron™) and rather than blocking all three dimensions, we will only block the outer two dimensions.

If the entire I1 line is brought into cache and we block on the outer two dimensions, we can hold 86 complete I1 lines of three variables in Level 2 cache. A blocking factor of 8 on both I2 and I3 would require 64 I1 lines in cache—this is a good conservative blocking factor.

Given these details, we decide to recode the loop as follows:

```
PARAMETER (BLOCK2 = 8, BLOCK3 = 8)
DO I3BLOCK = 2, N3-1,BLOCK3
DO I2BLOCK = 2, N2-1,BLOCK2
DO I3 = I3BLOCK, MIN(N3-1,I3BLOCK + BLOCK3-1)
   DO I2 = I2BLOCK, MIN(N2-1,I2BLOCK + BLOCK2-1)
      DO I1 = 1,N1
         R1(I1) = R(I1,I2-1,I3) + R(I1,I2 + 1,I3)
>              + R(I1,I2,I3-1) + R(I1,I2,I3 + 1)
         R2(I1) = R(I1,I2-1,I3-1) + R(I1,I2 + 1,I3-1)
>              + R(I1,I2-1,I3 + 1) + R(I1,I2 + 1,I3 + 1)
      ENDDO
      DO I1 = 2,N1-1
```

```
       U(I1,I2,I3) = U(I1,I2,I3)
>                   + C(0) * R(I1,I2,I3)
>                   + C(1) * (R(I1-1,I2,I3) + R(I1+1,I2,I3)
>                      + R1(I1))
>                   + C(2) * (R2(I1) + R1(I1-1) + R1(I1+1))
         ENDDO
       ENDDO
   ENDDO
   ENDDO
   ENDDO
```

By introducing blocking loops outside the original triple-nested DO loop, we control the blocking subgrid to fit into Level 2 cache. When this example is run on the quad-core Opteron, we obtain a 15% increase in performance. Other multicore systems have different size Level 2 and Level 3 caches, it is important to know these sizes to calculate the best block for your kernel. Blocking of the Level 3 cache would give larger block sizes; however, the performance is close to the value one gets by using this blocking.

6.2.5.1.2 Combining Multiple Loops   Another valuable cache blocking strategy is to restructure multinested DO loops into outerloops with multiple inner loops. For example, consider the following piece of code for the Leslie3D code, a computational fluid dynamics application. There are multiple restructurings that will improve this loop significantly. Earlier in this chapter, small loops were unrolled inside larger DO loops. Our first step will be to unroll NN inside of the K,J,I looping structure.

```
DO NN = 1,5
   DO K = K1,K2
      KK = K + KADD
      KBD = KK - KBDD
      KCD = KK + KBDD
      DO J = J1,J2
         DO I = I1,I2
            QAV(I,J,K,NN) = R6I * (2.0D0  * Q(I,J,KBD,NN,N) +
>                           5.0D0 * Q(I,J, KK,NN,N) -
>                                   Q(I,J,KCD,NN,N))
         END DO
      END DO
   END DO
```

```
END DO
DO K = K1,K2
   DO J = J1,J2
      DO I = I1,I2
         UAV(I,J,K) = QAV(I,J,K,2) / QAV(I,J,K,1)
      END DO
   END DO
END DO
DO K = K1,K2
   DO J = J1,J2
      DO I = I1,I2
         VAV(I,J,K) = QAV(I,J,K,3) / QAV(I,J,K,1)
      END DO
   END DO
END DO
DO K = K1,K2
   DO J = J1,J2
      DO I = I1,I2
         WAV(I,J,K) = QAV(I,J,K,4) / QAV(I,J,K,1)
      END DO
   END DO
END DO
DO K = K1,K2
   DO J = J1,J2
      DO I = I1,I2
         RKE = 0.5D0 * (UAV(I,J,K) * UAV(I,J,K) +
>                       VAV(I,J,K) * VAV(I,J,K) +
>                       WAV(I,J,K) * WAV(I,J,K))
         EI = QAV(I,J,K,5) / QAV(I,J,K,1) - RKE
         TAV(I,J,K) = (EI - HFAV(I,J,K,1)) / HFAV(I,J,K,3)
         PAV(I,J,K) = QAV(I,J,K,1) * HFAV(I,J,K,4) *
                                    TAV(I,J,K)
      END DO
   END DO
END DO
```

Following is the rewrite of the first quadruple-nested loop:

```
DO K = K1,K2
KK = K + KADD
   KBD = KK - KBDD
   KCD = KK + KBDD
```

```
DO J = J1,J2
   DO I = I1,I2
      QAV(I,J,K,1)= R6I * (2.0D0  * Q(I,J,KBD,1,N) +
>       5.0D0 * Q(I,J, KK,1,N) -
>                             Q(I,J,KCD,1,N))
      QAV(I,J,K,2)= R6I * (2.0D0  * Q(I,J,KBD,2,N) +
>       5.0D0 * Q(I,J, KK,2,N) -
>                             Q(I,J,KCD,2,N))
      QAV(I,J,K,3)= R6I * (2.0D0  * Q(I,J,KBD,3,N) +
>       5.0D0 * Q(I,J, KK,3,N) -
>                             Q(I,J,KCD,3,N))
      QAV(I,J,K,4)= R6I * (2.0D0  * Q(I,J,KBD,4,N) +
>       5.0D0 * Q(I,J, KK,4,N) -
>                             Q(I,J,KCD,4,N))
      QAV(I,J,K,5)= R6I * (2.0D0  * Q(I,J,KBD,5,N) +
>       5.0D0 * Q(I,J, KK,5,N) -
>                             Q(I,J,KCD,5,N))
      END DO
   END DO
END DO
```

Now we move the K and J loops to outside, leaving the I loop inside:

```
DO K = K1,K2
   KK = K + KADD
   KBD = KK - KBDD
   KCD = KK + KBDD
   DO J = J1,J2
     DO I = I1,I2
       QAV(I,J,K,1) = R6I * (2.0D0 * Q(I,J,KBD,1,N) +
>        5.0D0 * Q(I,J, KK,1,N) -
>                             Q(I,J,KCD,1,N))
       QAV(I,J,K,2) = R6I * (2.0D0 * Q(I,J,KBD,2,N) +
>        5.0D0 * Q(I,J, KK,2,N) -
>                             Q(I,J,KCD,2,N))
       QAV(I,J,K,3) = R6I * (2.0D0 * Q(I,J,KBD,3,N) +
>        5.0D0 * Q(I,J, KK,3,N) -
>                             Q(I,J,KCD,3,N))
       QAV(I,J,K,4) = R6I * (2.0D0 * Q(I,J,KBD,4,N) +
>        5.0D0 * Q(I,J, KK,4,N) -
>                             Q(I,J,KCD,4,N))
       QAV(I,J,K,5) = R6I * (2.0D0 * Q(I,J,KBD,5,N) +
```

```
>            5.0D0 * Q(I,J, KK,5,N) -
>                              Q(I,J,KCD,5,N))
       UAV(I,J,K) = QAV(I,J,K,2) / QAV(I,J,K,1)
       VAV(I,J,K) = QAV(I,J,K,3) / QAV(I,J,K,1)
       WAV(I,J,K) = QAV(I,J,K,4) / QAV(I,J,K,1)
       RKE = 0.5D0 * (UAV(I,J,K) * UAV(I,J,K) +
>          VAV(I,J,K) * VAV(I,J,K) +
>          WAV(I,J,K) * WAV(I,J,K))
       EI = QAV(I,J,K,5) / QAV(I,J,K,1) - RKE
       TAV(I,J,K) = (EI - HFAV(I,J,K,1)) / HFAV(I,J,K,3)
       PAV(I,J,K) = QAV(I,J,K,1) * HFAV(I,J,K,4)
                     * TAV(I,J,K)
      END DO
    END DO
END DO
```

Several beneficial optimizations have been performed in this loop. One major optimization is that we have been able to eliminate several divides, by merging all the I loops together. This results in a savings in having to perform the 1/QAV(I,J,K,1) only once instead of three times in the original. This is due to the compiler replacing the divide with a reciprocal approximation, 1/QAV(I,J,K,1) and then performing the divides by multiplying this by the numerator. This particular optimization may lose some precision and can be inhibited with appropriate compiler flags.

For another example of loop reordering, revisit the example from the Parallel Ocean Program (POP) in Chapter 3. There, Fortran 90 array syntax was replaced with a single triple-nested DO loop, achieving a factor of two in speedup.

6.2.5.1.3 Elimination of Fetches and Stores   Floating point operations are becoming free compared to memory fetching/storing. Consider the MGRID example earlier in this section:

```
PARAMETER (BLOCK2 = 8, BLOCK3 = 8)
DO I3BLOCK = 2,N3-1,BLOCK3
DO I2BLOCK = 2,N2-1,BLOCK2
DO I3 = I3BLOCK,MIN(N3-1,I3BLOCK+BLOCK3-1)
   DO I2 = I2BLOCK,MIN(N2-1,I2BLOCK+BLOCK2-1)
     DO I1 = 1,N1
        R1(I1) = R(I1,I2-1,I3) + R(I1,I2+1,I3)
>          + R(I1,I2,I3-1) + R(I1,I2,I3+1)
        R2(I1) = R(I1,I2-1,I3-1) + R(I1,I2+1,I3-1)
```

```
>                +R(I1,I2-1,I3+1) +R(I1,I2+1,I3+1)
      ENDDO
      DO I1 = 2,N1-1
         U(I1,I2,I3) =U(I1,I2,I3)
>            +C(0) * R(I1,I2,I3)
>            +C(1) * (R(I1-1,I2,I3) +R(I1+1,I2,I3)
>                 +R1(I1))
>            +C(2) * (R2(I1) +R1(I1-1) +R1(I1+1))
      ENDDO
   ENDDO
ENDDO
ENDDO
ENDDO
```

Several memory loads and stores can be eliminated by combining the two innermost loops on I1. The issue is that more computation must be performed for the variables required in the second DO loop to be available. A potential rewrite is

```
DO I3BLOCK = 2,N3-1,BLOCK3
DO I2BLOCK = 2,N2-1,BLOCK2
DO I3 = I3BLOCK,MIN(N3-1,I3BLOCK+BLOCK3-1)
   DO I2 = I2BLOCK,MIN(N2-1,I2BLOCK+BLOCK2-1)
     DO I1 = 2,N1-1
        R1I1 = R(I1,I2-1,I3) + R(I1,I2+1,I3)
>           +R(I1,I2,I3-1) + R(I1,I2,I3+1)
        R2I1 = R(I1,I2-1,I3-1) + R(I1,I2+1,I3-1)
>           +R(I1,I2-1,I3+1) + R(I1,I2+1,I3+1)
        R1I1M1 = R(I1-1,I2-1,I3) + R(I1-1,I2+1,I3)
>            +R(I1-1,I2,I3-1) + R(I1-1,I2,I3+1)
        R1I1P1 = R(I1+1,I2-1,I3) + R(I1+1,I2+1,I3)
>            +R(I1+1,I2,I3-1) + R(I1+1,I2,I3+1)
        U(I1,I2,I3) =U(I1,I2,I3)
>           +C(0) * R(I1,I2,I3)
>           +C(1) * ( R(I1-1,I2,I3) + R(I1+1,I2,I3)
>                 +R1I1 )
>           +C(2) * ( R2I1 + R1I1M1 + R1I1P1 )
      ENDDO
   ENDDO
ENDDO
ENDDO
ENDDO
```

Interestingly, this restructuring runs about 10% faster even though it introduces six more additions. Although the rewrite only gets 10% improvement, it does illustrate an important technique when the memory bandwidth is the bottleneck. Some memory-limited regions of code may be rewritten to recalculate some of the operands rather than storing and retrieving those values from memory. Of course, the performance would vary on a case by case basis.

6.2.5.1.4 Variable Analysis in Multiple DO Loops   When multiple DO loops are present in an important kernel, benefit may be obtained by counting the amount of data used in the loops to see if strip mining can be used to reduce the memory footprint and obtain better cache performance. Consider the following simple example that illustrates a very powerful restructuring technique for improving cache performance. Note that in the first triple-nested DO loop, we used NZ * NY * NX elements of the A array. Then in the second triple-nested DO loop, the same array is used. Now this is a silly example, since it first multiples A by 2.0 and then multiplies the result by 0.5. However, imagine multiple DO loops that do reasonable computation. How much data are loaded into cache in the first multiple-nested loop and then are that data still around when we get to the second multinested loop?

```
DO iz = 1,nz
DO iy = 1,ny
DO ix = 1,nx
   a(ix,iy,iz) = a(ix,iy,iz) * 2.0
END DO
END DO
END DO
DO iz = 1,nz/in
DO iy = 1,ny
DO ix = 1,nx
   a(ix,iy,iz) = a(ix,iy,iz) * 0.5
END DO
END DO
END DO
```

If NX = NY = 100 and NZ = 512, the total amount of data accessed in each loop is 5,120,000 words. If A is 8 bytes, then 40,960,000 bytes are used. This is much larger than the caches in the multicore chip and the second loop will have to refetch everything from memory. Now consider the following rewrite that introduces a strip mining loop around both multinested DO loops. This loop will only update a segment of the A array and if IC is large enough, the portion of the A array that is updated may fit into the cache.

```
DO ic = 1,nc*in
    DO iz = 1,nz/in
    DO iy = 1,ny
    DO ix = 1,nx
       a(ix,iy,iz) = a(ix,iy,iz) * 2.0
    END DO
    END DO
    END DO
    DO iz = 1,nz/in
    DO iy = 1,ny
    DO ix = 1,nx
       a(ix,iy,iz) = a(ix,iy,iz) * 0.5
    END DO
    END DO
    END DO
END DO
```

The following table gives the amount of data that is accessed in the two loops:

NX	NY	NZ	IC	M Words	MB	Level 1 Refills (65,526)	Level 2 Refills (524,288)	Level 3 Refills (2 MB)
100	100	512	1	5.12	40.96	635	80	20
100	100	256	2	2.56	20.48	318	40	10
100	100	128	4	1.28	10.24	159	20	5
100	100	64	8	0.64	5.12	79	10	2.5
100	100	32	16	0.32	2.56	39	5	1.25
100	100	16	32	0.16	1.28	19	2.5	0.625
100	100	8	64	0.08	0.64	9	1.28	0.3675
100	100	4	128	0.04	0.32	4.5	0.64	0.1837
100	100	2	256	0.02	0.16	2.25	0.32	0.0918

From this table, the performance of this multinested DO loop should improve dramatically when IC is 16 and then even improve more when

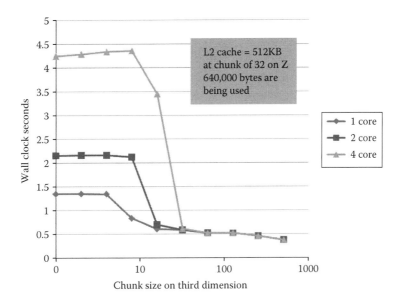

FIGURE 6.24 Comparisons of multiple runs using different cache blocking chunks.

IC is 128 and finally run best when IC is 256. To illustrate the dramatic impact this has on the memory utilization of this loop, we run this example packed. That is, we run four copies of this loop at the same time on a quad-core socket and measure the overall runtime for this example. Figure 6.24 gives the time the kernel uses as a function of IC.

Note that the unpacked version (1 core series) that runs one of the multicores starts improving when IC is greater than 8. All of the copies of the kernels finish at the same time at an IC of 32 when the array section fits into Level 3 cache. Then, as IC increases, the time for all the kernels to finish starts improving. This indicates that there is enough bandwidth for the multicore socket when the size of the data set fits into Level 3 memory and then as the working set fits into Level 2 memory the time reduces more. This is an extremely important analysis that application developers should consider when designing the loop structures.

6.2.5.1.5 *Accessing Data with Strides and Indirect Addressing*   Earlier in this chapter, we discussed the performance of different memory access patterns. Strides and indirect addressing can significantly hurt performance. While these access patterns can be vectorized, they lose more performance because they can end up fetching much more data from memory than they use. In Chapter 2, the details of the cache line were discussed which indicated that any memory access would result in a full cache line of 64 bytes being loaded from memory to Level 1 cache. When an application strides through this cache line, it does not utilize all the data in the cache line. When using 8-byte reals, the cache line contains eight data elements. If the stride is 2, only 4 of the elements will be used, as the stride becomes larger than 8, the number of elements used per cache line fetched to the cache becomes 1. So 1/8 of the data fetched from memory would be used, thus reducing the effective memory bandwidth by a factor of 8.

With indirect addressing, the situation is less clear. For example, consider an equation of state routine in a finite-difference code. Frequently, there will be variables dimensioned on the material contained within the grid block—Density(MAT(I,J,K)). This might give the relative density of the material that is contained in grid block I,J,K and it is definitely indirect addressing. The difference in this situation is that frequently adjacent grid blocks may contain the same material. If I + 1,J,K grid block contains the same material, then Density(MAT(I + 1,J,K)) is the same memory location as Density(MAT(I,J,K)); hence, we actually get cache reuse.

Another example where indirect addressing frequently gives good cache locality is in a compressed sparse vector where the indices given by the indirect address are actually contiguous in memory for a given row of a matrix.

When indirect addressing is totally random and does not point to contiguous addresses, performance suffers accordingly. An interesting example of optimization of such a case was in a Particle in Cell code where the indirect addresses are sorted to give a more uniform memory access pattern. If the order of processing the particles is not important, then spending some extra time sorting the particle list may result in more efficient memory addressing and give an overall performance gain.

6.3 SUMMARY

In this chapter, we have used a wide variety of restructuring techniques. While the examples themselves may not be similar to important computational kernels in your application, the techniques should be understood so that they can be applied to nonvectorizable constructs.

6.3.1 Loop Reordering

This is a very powerful restructuring technique that is seldom used by the compiler due to the complication with data dependency analysis. Loop reordering can be used to restructure a loop nest to get better array accessing, to vectorize on the longer index, and so on. Whenever there are statements between the loops being switched, those loops must be split out from the subsequent multinested loop (check out the restructuring of DO 46030 and DO 45020).

6.3.2 Index Reordering

Index reordering is used to remove strides on arrays. This must be used with caution, because it is a global change. When the indices of an array are reordered, the restructuring has global implications. If the array is passed as an argument to a subroutine, both the caller and the callee must be modified. If the array is in a module or a common block, all the routines that access that module or common block must be rewritten.

6.3.3 Loop Unrolling

Loop unrolling is a very powerful technique to increase computational intensity in a DO loop. Additionally, loop unrolling gives the compiler more operations to shuffle around and get more overlap. This technique is often employed by the compiler; however, the user should not be afraid of using it.

6.3.4 Loop Splitting (Loop Fission)

Loop splitting is used to remove recursion, subroutine calls, and nonvectorizable constructs from loops. Additionally, when other restructuring techniques, such as loop switching, are employed, frequent loop splitting is needed to enable the isolation of statement between loops. On multicore systems with SSE instructions, loop splitting introduces significant overhead and may not be beneficial.

6.3.5 Scalar Promotion

Scalar promotion is needed when a loop is split and a scalar contains information for each value of the loop index. In this case, the scalar must be promoted to an array. Scalar promotion is used extensively in the restructuring of DO loop 47020. When scalar promotion is used, it is important to keep the array sizes small enough to fit into the cache. If the promoted scalar does not fit into the cache, then the overhead of promoting it may steal any performance that is gained in the restructuring.

6.3.6 Removal of Loop-Independent IFs

In DO loop 47020, the inner I loop was split up into numerous loops on I and the loop-independent IF statements were removed from the inner DO loops. This is very powerful and another restructuring technique that is not typically employed by the compiler. This restructuring may require the introduction of arrays of promoted scalars; however, as in DO loop 47020, these were kept small to be cache resident.

6.3.7 Use of Intrinsics to Remove IFs

Whenever MAX, MIN, ABS, and SIGN can be used to replace an IF statement, a significant performance gain can be achieved. The compiler converts these intrinsics into very efficient machine operations.

6.3.8 Strip Mining

In the rewrite of DO loop 47120, strip mining was used to reduce the number of operations performed in each pass through the inner DO loop. In this case, strip mining was used to reduce the number of unnecessary operations; however, another very powerful strip mining technique is to reduce the amount of memory used when promoting scalars to arrays. By having a strip-mining loop around a set of loops where scalar temporaries must be introduced, the total amount of data allocated can be reduced to fit better into the cache.

6.3.9 Subroutine Inlining

Subroutine inlining is used extensively in compilers. Given the powerful directives compilers supply for inlining, the user may not have to perform their own inlining. This is particularly a very important optimization for C and C++ applications. By inlining a set of routines, the compiler has a much better chance of resolving potential data dependencies.

6.3.10 Pulling Loops into Subroutines

Often an application will call subroutines to perform minor operations such as add or multiply. Anytime very simple operations are contained within a subroutine, the overhead for calling the subroutine may be more than the computation. A simple restructuring that helps if the small routine is called from within DO loops is to create an array-valued subroutine. For example, replace CALL ADD(A,B) with CALL VADD(N,A,B). Now if N is very large, the overhead of calling the routine is minimized.

The rule is to never make a subroutine call to perform any operations that take less time than the time to call the subroutine.

6.3.11 Cache Blocking

Cache blocking is the best technique for obtaining better cache utilization for multinested loops. The example in this chapter is an excellent template to follow for blocking differencing loops.

6.3.12 Loop Jamming (Loop Fusion)

The restructuring performed in Chapter 3, rewriting Fortran 10 syntax to a DO loop with several statements, is a good example of loop jamming. Additionally, the example in Section 6.2.5.1.2 shows how one can benefit from jamming many DO loops together.

EXERCISES

1. You should time many of these examples yourself on the machine that is your major workhorse. It is very important to understand that a few of the timings were performed on operands that originate from memory. How can one assure that the timing of the example is performed on operands coming from memory? On the other hand, how can one be assured that the operands come from the cache?
2. How do array dimensioning and DO loop nesting affect memory performance?
3. How do array indexing strides affect TLB and cache utilization?
4. What is computational intensity?
   a. In the following DO loop, why do not we count C(J) as a load? What is the computational intensity of this loop?

   ```
   DO I = 1, N
      A(I) = B(I) * C(J)
   ENDDO
   ```

5. How can loop unrolling be used to improve the performance of matrix multiply?
6. What is packed and unpacked execution? How do they relate to timing runs?
7. How do subroutine and function calls affect the performance of loops? How can loops with function and subroutine calls be restructured to improve performance?

8. How might vector length affect the decision to use library BLAS routines?

9. How do vector length and stride interact in optimizing loop nests?

10. What is loop blocking? How can it be used to improve performance?

11. How can additional operations be used to reduce memory accesses? Why is this useful?

12. There is a function in Fortran called LOC (or LOCF on some systems), which returns the physical address of the argument of an array. Investigate the storage of some of the major arrays in your application. Is memory allocated at compile time more likely to be contiguous in physical memory than if the memory is allocated with malloc and/or allocate during run-time. Investigate the layout of the physical memory when large chunks of memory are allocated using malloc and/or allocate versus numerous smaller chunks of memory. Which approach generates more fragmented memory?

13. One way of increasing the computational intensity of a multi-level DO loop is by unrolling an outer DO loop inside an inner DO loop. This is not always helpful. Come up with an example where it will work and an example where it will not work.

14. If one uses a system with multicore sockets, one should rerun the previous assignments in packed mode; that is, if on a four-core socket, run four copies of the test at the same time.

15. When restructuring DO loops for GPGPUs, which can deliver much higher performance gains than SSE3 systems, what are the main restructuring techniques that can be used even though they introduce overheads?

16. What restructuring techniques can be performed by using directives or compiler flags and what will probably have to be performed by hand?

17. When restructuring an application it is important to have a performance portable version of the application. That is, a program that performs well on any architecture that it is run on. What are some of the areas where a restructuring for a GPGPU might not run well on the multicore socket? How might such situations be handled?

CHAPTER 7

Parallelism across the Nodes

IN THIS CHAPTER, WE LOOK AT how to improve the scalability of applications. In some sense, this is the most difficult challenge for the programmer. There are many different reasons why an application might not scale and some of those may be impossible to improve without a total rewrite of the application. In order to hit as many of the scenarios as possible, we look at numerous applications and several different MPI implementations and discuss the reasons why they scale and why they eventually reach their limits of scalability. What are the principal hazards to scalability? The most prevalent is the lack of sufficient work. When a fixed problem is scaled to a large number of processors, the application will sooner or later quit scaling. The individual processors will run out of work to do and communication will dominate. The first code we will examine, one of the SPEC® MPI benchmark suites, is a case of trying to run a small problem on too many processors. Other hazards are less obvious and sometimes no matter how big a problem is, the application does not scale. For example:

1. Having a poor decomposition that limits the number of processors that can be effectively used.

2. Having an application that is dependent on a nonscalable type of MPI message passing. The typical MPI_ALLREDUCE or MPI_ALLTOALL is not a scalable construct. As the number of processors increase, either the number of messages (in a strong scaling situation)

of the amount of data (in a weak scaling situation) increases nonlinearly and communication will soon dominate.

3. Having severe load imbalance. This imbalance can be due to computation and/or communication.

4. When scaling nearest neighbor communication on an interconnect with insufficient link bandwidth, poor MPI task placement impacts scaling.

These are a few of the most common hazards to scaling.

7.1 LESLIE3D

Leslie3D is a "large Eddy simulation" code that performs finite differences on a 3D grid. Leslie3D uses 3D decomposition so that the mesh is divided among the processors as small cubes.

Figure 7.1 is a scaling chart for running the application up to 2048 cores.

We see that the scaling departs from linear at around 512 processors. Looking at the scaling of individual routines reveals a couple of potential scaling bottlenecks. This is an example of strong scaling; that is, the problem size stays the same and as the number of processors increase, the work per processor decreases (Figure 7.2).

In fact, all the computational routines are obtaining superlinear speedup. They attain a factor of as high as 51 going from 64 to 2048 cores. This is what is termed "superlinear scaling," as the processor count increases, the increase in performance is larger. In this case, one of the

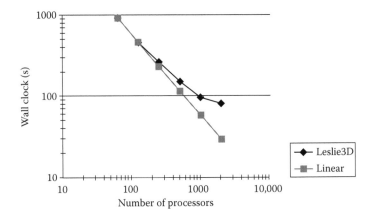

FIGURE 7.1 Scaling of Leslie3D.

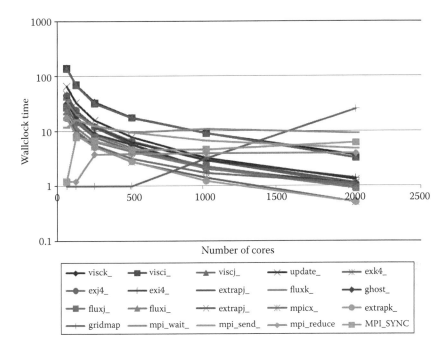

FIGURE 7.2   Scaling of individual routines from Leslie3D.

routines increased by a factor of 51 when the processor count increased by 32. The reason for this phenomenon is that as the working set on the processor becomes smaller, it fits into cache and array references that require actual memory transfers are reduced.

So, the computation is scaling extremely well and the load balance of the application is good, and it never really changes as we scale up. One large inefficiency occurs in a grid map at the higher processor counts. Fortunately, this routine is only called during initialization and if we run the computation to larger time steps, the time does not grow. Unfortunately, the grid generation for a run at 4096 cores took almost an hour to complete. There are cases where initialization is exponential in time versus processor count. For very large problem sizes, we may have to worry about the time to generate the grid. In this particular case, the grid partitioning is not parallelized. There are many parallel grid partitioning algorithms; however, their value is very problem dependent.

If we ignore the initialization, the other issue impacting the scaling is the MPI routines. While the time they take does not increase significantly with the core count, the time they take also does not decrease and so the overall percentage of total time increases. The MPI calls start to dominate at larger processor counts. At 64 processors, MPI takes 8.7% of the total time; at 2048, they take 26% of the time. The actual time taken by the MPI routines reduce from 50 seconds on 64 processors to 22 seconds on 2048 processors. Since the communication does not scale as well as the computation, it quickly becomes the bottleneck.

Given these statistics, we can conclude that the problem size, which is fixed in this case, is becoming too small. At 2048 processors, each processor has a cube of 15 on a side. This is definitely a case where we need a larger problem.

7.2 PARALLEL OCEAN PROGRAM (POP)

The next application that we examine is Parallel Ocean Program (POP) [7], originally developed for the connection machine. POP is widely used in the climate community and has some interesting scaling characteristics. POP consists of two major computational parts. First, the BAROCLINIC portion computes the differencing equations on the 3D grid. POP is decomposed on longitude and latitude dimensions and the third dimension is depth. The second compute portion is the BAROCLINIC where a conjugate gradient solver works on the 2D surface of the ocean. The data set used is the 1/10 grid which is the largest production problem available at this time. This is another example of strong scaling.

While the BAROCLINIC portion is extremely scalable, in fact, it scales superlinearly. The BAROTROPIC is comprised of numerous global sums which require significant processor synchronization. The timings for STEP include the entire time step loop which includes BAROCLINIC and BAROTROPIC.

	Wallclock Seconds			Scaling		
Processors	Step	Baroclinic	Barotropic	Step	Baroclinic	Barotropic
600	462.79	321.87	140.92	1.00	1.00	1.00
1200	208.11	125.58	82.53	2.22	2.56	1.71
2400	143.99	62.52	81.47	3.21	5.15	1.73

An example from POP was used in Chapter 3 to show the inefficiencies of Fortran 90 array syntax and this inefficiency goes away once the subgrid on the processor gets small enough to fit into cache. This is the reason for the superlinear improvement in the Baroclinic portion. The BAROTROPIC section consists of an iteration loop that performs a nine-point stencil and two to three global sums for each iteration. This time is dominated by the boundary exchange in the nine-point stencil and the global reductions. Following is a profile showing the details of the largest run:

```
Time %  | Time       | Imb. Time | Imb.    | Calls     |Experiment = 1
        |            |           | Time %  |           |Group
        |            |           |         |           | Function
        |            |           |         |           | PE = 'HIDE'
100.0%  |179.022039 |       --  |    --   |2831065.2  |Total
|--------------------------------------------------------------
|34.5%  | 61.781441 |       --  |    --   |167631.0   |MPI_SYNC
||-------------------------------------------------------------
||31.6% | 56.622371 |83.976148 |59.8%    |155081.0   |mpi_allreduce_(sync)
||2.4%  |  4.350423 | 1.963490 |31.1%    |  7475.0   |mpi_bcast_(sync)
||=============================================================
|33.9%  | 60.608345 |       --  |    --   |790019.8   |USER
||-------------------------------------------------------------
Computational routines removed.
||=============================================================
|31.5%  | 56.414324 |       --  |    --   |1649969.4  |MPI
||-------------------------------------------------------------
||13.3% | 23.783400 | 1.523985 | 6.0%    | 155081.0  |mpi_allreduce_
||13.2% | 23.573875 |34.661049 |59.5%    | 585029.7  |mpi_waitall_
||2.0%  |  3.548423 | 7.234560 |67.1%    | 448646.9  |mpi_irecv_
||1.4%  |  2.522932 | 0.193564 | 7.1%    |   7475.0  |mpi_bcast_
||1.3%  |  2.381773 | 5.676421 |70.5%    | 448637.2  |mpi_isend_
|=============================================================
```

Several lessons are to be learned here. Note that the top time consumer is SYNC time. This is primarily the time that processors wait to do the global sum, which is performed with the MPI_ALLREDUCE. Next is the MPI_WAITALL which is where the processors wait for the halo exchange in the nine-point stencil in the BAROTROPIC region. So at 2400 processors, MPI SYNC time and MPI use 66% of the wall-clock time. Note that we have a tremendous amount of load imbalance. This load imbalance must be caused by either the computation or some communication. While the load imbalance shows up in the calls to MPI_WAITALL and the MPI_ALLREDUCE synchronization time, the load imbalance cannot be attributed to the MPI_ALLREDUCE. The MPI_ALLREDUCE sync is gathered by inserting instrumentation at a barrier prior to the MPI_ALLREDUCE call to identify any load imbalance that existed prior to the MPI_ALLREDUCE call. So 31% of the time to execute this code is due to waiting prior to the MPI_ALLREDUCE call. If we can identify and fix this load imbalance we will be able to improve the performance of the application significantly. If we examine the operations that are performed prior to the MPI_ALLREDUCE, we find the nine-point stencil which is performed for all of the processors and the amount of communication does not vary from processor to processor. The only other operation that might be causing load imbalance is the nearest neighbor communication. While the communication is the same for each processor, there may be an issue with the way the communication is performed that causes the load imbalance.

Consider the following excerpt of codes from the POP ocean model:

```
!-----------------------------------------------------------------
  do j = jphys_b,jphys_e
    do i = iphys_b,iphys_e
      XOUT(i,j) = CC(i,j)*X(i, j) +
  &               CN(i,j)*X(i, j+1)+CN(i,j-1)*X(i, j-1) +
  &               CE(i,j)*X(i+1,j)+CE(i-1,j)*X(i-1,j) +
  &               CNE(i,j)*X(i+1,j+1)+CNE(i,j-1)*X(i+1,j-1) +
  &               CNE(i-1,j)*X(i-1,j+1)+CNE(i-1,j-1)*X(i-1,j-1)
    end do
  end do
!-----------------------------------------------------------------
!
! update ghost cell boundaries.
!
!-----------------------------------------------------------------
  call MPI_IRECV(XOUT(1,1), 1, mpi_ew_type, nbr_west,
  &            mpitag_wshift, MPI_COMM_OCN, request(3), ierr)
  call MPI_IRECV(XOUT(iphys_e+1,1), 1, mpi_ew_type, nbr_east,
  &            mpitag_eshift, MPI_COMM_OCN, request(4), ierr)
  call MPI_ISEND(XOUT(iphys_e+1-num_ghost_cells,1), 1,
  &            mpi_ew_type, nbr_east,
  &            mpitag_wshift, MPI_COMM_OCN, request(1), ierr)
  call MPI_ISEND(XOUT(iphys_b,1), 1, mpi_ew_type, nbr_west,
  &            mpitag_eshift, MPI_COMM_OCN, request(2), ierr)
  call MPI_WAITALL(4, request, status, ierr)
  call MPI_IRECV(XOUT(1,jphys_e+1), 1, mpi_ns_type, nbr_north,
  &            mpitag_nshift, MPI_COMM_OCN, request(3), ierr)
  call MPI_IRECV(XOUT(1,1), 1, mpi_ns_type, nbr_south,
  &            mpitag_sshift, MPI_COMM_OCN, request(4), ierr)
  call MPI_ISEND(XOUT(1,jphys_b), 1, mpi_ns_type, nbr_south,
  &            mpitag_nshift, MPI_COMM_OCN, request(1), ierr)
```

```
call MPI_ISEND(XOUT(1,jphys_e+1-num_ghost_cells), 1,
&          mpi_ns_type, nbr_north,
&          mpitag_sshift, MPI_COMM_OCN, request(2), ierr)
call MPI_WAITALL(4, request, status, ierr)
```

This is very poor MPI point-to-point message passing. In this sequence of code, the XOUT array is computed; then nonblocking receives and sends are posted. This does effectively overlap the send/recv to the east and west neighbor; however, there is no assurance that the receives are posted before the messages from the neighbors arrive. There is then a MPI_WAITALL to wait until the east/west messaging is completed. Then the north/south messages are posted, once again overlapping the two sends, however, not assuring that the receives are preposted prior to the messages arriving. As discussed earlier in Chapter 3, when receives are not preposted the receiving processor must allocate a buffer to hold the message until the receive is posted. Until the receive is posted, the processor has no idea where to put the message. In a run of 2048 processors, some of the receives may be preposted while other receives are not preposted. When the receive is preposted and the processor can move the message directly into the application array, the time it takes to receive the message will be faster. If, on the other hand, the receive is not preposted, the time to allocate the buffer receive the message and then move it into the application buffer takes significantly longer. Not preposting the receives is introducing load-imbalance into the application.

Consider the following rewrite preformed by Pat Worley of Oak Ridge National Laboratory [12]:

```
call MPI_IRECV(XOUT(1,1), 1, mpi_ew_type, nbr_west,
&              mpitag_wshift, MPI_COMM_OCN, request(1), ierr)
call MPI_IRECV(XOUT(iphys_e+1,1), 1, mpi_ew_type, nbr_east,
&              mpitag_eshift, MPI_COMM_OCN, request(2), ierr)
call MPI_IRECV(XOUT(1,jphys_e+1), 1, mpi_ns_type, nbr_north,
&              mpitag_nshift, MPI_COMM_OCN, request(5), ierr)
call MPI_IRECV(XOUT(1,1), 1, mpi_ns_type, nbr_south,
&              mpitag_sshift, MPI_COMM_OCN, request(6), ierr)
!-------------------------------------------------------------
do j = jphys_b,jphys_e
  do i = iphys_b,iphys_e
    XOUT(i,j) = CC(i,j)*X(i, j) +
&              CN(i,j)*X(i, j+1) + CN(i,j-1)*X(i, j-1) +
```

```
&                    CE(i,j)*X(i+1,j ) +CE(i-1,j)*X(i-1,j ) +
&                    CNE(i,j)*X(i+1,j+1) +CNE(i,j-1)*X(i+1,j-1) +
&                    CNE(i-1,j)*X(i-1,j+1) +CNE(i-1,j-1)*X(i-1,j-1)
  end do
end do
i=1
do n=1,num_ghost_cells
  do j=jphys_b,jphys_e
    buffer_east_snd(i) =XOUT(iphys_e+n-num_ghost_cells,j)
    buffer_west_snd(i) =XOUT(iphys_b+n-1,j)
    i=i+1
  end do
end do
call MPI_ISEND(buffer_east_snd,
&              buf_len_ew, MPI_DOUBLE_PRECISION, nbr_east,
&              mpitag_wshift, MPI_COMM_OCN, request(3), ierr)
call MPI_ISEND(buffer_west_snd,
&              buf_len_ew, MPI_DOUBLE_PRECISION, nbr_west,
&              mpitag_eshift, MPI_COMM_OCN, request(4), ierr)
call MPI_WAITALL(2, request, status_wait, ierr)
i=1
do n=1,num_ghost_cells
  do j=jphys_b,jphys_e
    XOUT(n,j)=buffer_west_rcv(i)
    XOUT(iphys_e+n,j) =buffer_east_rcv(i)
    i=i+1
  end do
end do
!-----------------------------------------------------------------
! send north-south boundary info
!-----------------------------------------------------------------
call MPI_ISEND(XOUT(1,jphys_e+1-num_ghost_cells), buf_len_ns,
&              MPI_DOUBLE_PRECISION, nbr_north,
&              mpitag_sshift, MPI_COMM_OCN, request(1), ierr)
call MPI_ISEND(XOUT(1,jphys_b), buf_len_ns,
&              MPI_DOUBLE_PRECISION, nbr_south,
&              mpitag_nshift, MPI_COMM_OCN, request(2), ierr)
call MPI_WAITALL(6, request, status, ierr)
```

In this code, at the very outset, four receives are preposted. Each of these receives pass an array to hold the incoming message. These arrays are not in use and are not needed until the message is received. Now the nine-point stencil is performed by each processor and once this computation is complete, the buffers for the east–west messages are packed into a contiguous section of memory. Once this is completed, the sends are made to the east–west neighbor processors. At this time, the processor must wait for two of the receives to be completed. Now the east–west messages are unpacked into the XOUT array. This is done in order to assure that the corners are available to be sent to the north and south neighbors. Finally, the sends are made to the north and south neighbors and all processors must wait for all messages to be completed.

This rewrite has two major advantages. First, we have overlapped the communication with the computation; second, the likelihood that the receives are posted prior to the message arriving is very high. This will significantly improve the load imbalance and the overall performance of these point-to-point message calls. Using the above rewrite, the scaling of POP has been dramatically improved.

7.3 SWIM

The next application is a simple 2D benchmark that performs the shallow water equations. We will revisit this application in the next chapter to see how it might be improved using a hybrid approach; that is, using OpenMP™ on the node and MPI between nodes. Figure 7.3 is a scaling curve from this simple benchmark.

In examining the code, which originated from a testing version, a bad inefficiency was discovered that does not impact scaling; however, the

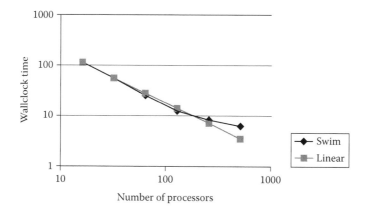

FIGURE 7.3   Scaling of the SWIM benchmark.

single node performance is poor. Following is hardware statistics from one of the major computational routines:

```
===============================================================
USER / calc3_
---------------------------------------------------------------
Time%                                   15.1%
Time                                    1.705979 secs
Imb.Time                                0.055310 secs
Imb.Time%                               3.2%
Calls                    701.3/sec      1198.0 calls
PAPI_L1_DCM              9.804 M/sec     16748308 misses
PAPI_TLB_DM             8.691 M/sec      14846920 misses
PAPI_L1_DCA             18.303 M/sec     31265669 refs
PAPI_FP_OPS                             0 ops
User time (approx)       1.708 secs      4099826899 cycles  100.0% Time
Average Time per Call    0.001424 sec
CrayPat Overhead : Time  0.1%
HW FP Ops / User time                   0 ops            0.0% peak(DP)
HW FP Ops / WCT
Computational intensity  0.00 ops/cycle  0.00 ops/ref
MFLOPS (aggregate)       0.00M/sec
TLB utilization          2.11 refs/miss  0.004 avg uses
D1 cache hit,miss ratios  46.4% hits     53.6% misses
D1 cache utilization (M)  1.87 refs/miss  0.233 avg uses
===============================================================
```

Note the horrendous TLB utilization: only 2.11 references per TLB miss. As discussed earlier in this book, we would like to see something closer to 500. In examining CALC3, we discover the reason for this inefficiency.

```
      DO 300 I = 1, M
      DO 300 J = js, je
      UOLD(I,J) = U(I,J) + ALPHA* (UNEW(I,J) - 2.*U(I,J) + UOLD(I,J))
      VOLD(I,J) = V(I,J) + ALPHA* (VNEW(I,J) - 2.*V(I,J) + VOLD(I,J))
      POLD(I,J) = P(I,J) + ALPHA* (PNEW(I,J) - 2.*P(I,J) + POLD(I,J))
      U(I,J) = UNEW(I,J)
      V(I,J) = VNEW(I,J)
      P(I,J) = PNEW(I,J)
300   CONTINUE
```

Notice that the J loop is on the inside. It is interesting that the compiler did not automatically invert the I and J loops. This was not coded in this way in the original version of SWIM; however, this is a nice example to illustrate how processor inefficiencies may potentially be scaled away. As the processor count increases, the size of the subgrid being computed becomes smaller and the TLB and Level 1 cache performance actually improves. MPI implementation is a form of high-level cache blocking. If we fix this inefficiency, we see that we improve the performance of the code significantly for low processor counts and less at high processor counts.

	Swim-Opt	Swim	Improvement
16	62.11	113	1.82
32	33	56	1.70
64	19	25	1.32
128	11.93	12.47	1.05
256	7.99	8.13	1.02
512	6.27	6.54	1.04

So while this improves the code significantly at low processor counts, it does not fix the scaling at high processor counts. So the major reason for the lack of scaling is the size of the problem.

This example shows an obscure reasoning: Why optimize single processor performance, unless it is obvious as in this example, until one performs the scaling study? In the next chapter, we will investigate this further.

7.4 S3D

S3D is a 3D combustion application that exhibits very good weak scaling. In setting up the input for S3D, the user specifies the size of the computational gird per processor. As the number of processors increases, the number of grid blocks increases accordingly. The communication in S3D is all nearest neighbors. Although S3D is a computational fluid dynamics (CFD) code, it is not a spectral model, it solves the physical equations using a higher order Runge–Kutta algorithm which does not require global transposes as the spectral model does. In Chapter 5, a weak scaling study of S3D was given that showed considerable increase in the cost of the computation (Figure 7.4).

Note in Figure 7.4 that the cost per grid point per time step stays constant as the number of processors grows. In this case, we see a significant departure from perfect scaling in going from 12,000 to 48,000 processors. To understand the deviation from good scaling, we must understand the

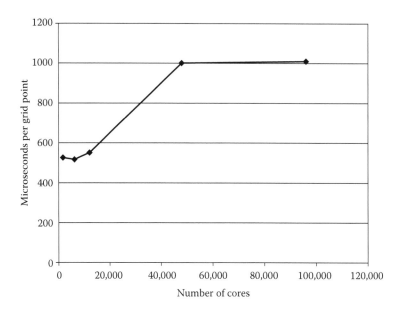

FIGURE 7.4   Weak scaling study of S3D.

type of communication that is performed within the S3D application. The computational scheme uses a fourth-order Runge–Kutta algorithm, which requires nearest-neighbor communication in all three directions. At 48,000 processors, the processor grid is $30 \times 40 \times 40$ and each processor contains a subgrid, $20 \times 20 \times 20$. When the differencing is performed in the first (x) dimension, each processor communicates with its neighbors on its right and left sides (Figure 7.5).

When the differentiation is performed in the second (y) dimension, each processor communicates with its neighbor above and below (Figure 7.6).

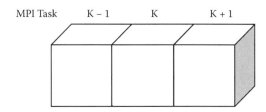

FIGURE 7.5   Communication in x direction.

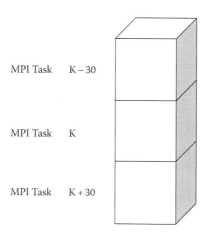

FIGURE 7.6   Communication in the y dimension.

And finally, when the differentiation is performed in the third (z) dimension, each processor communicates with its neighbor behind it and in front of it. Those processors are separated by 1200 processors (30 × 40) (Figure 7.7).

In a multicore MPP system, the default would be to sequentially place the MPI tasks on a node. So in a 12-core multicore node, MPI Task 1–12 would be on the first node, 13–24 on the second, and so on (Figure 7.8).

This is not the optimal alignment for a 3D code like S3D. Ideally, one would like to have 1,2,3,31,32,33,1201,1202,1203,1231,1232,1233 on the first node and then 4,5,6,34,35,36,1204,1205,1206,1234,1235,1236, and so on. By grouping these 3,2,2 3D quadrahedrals, the majority of the communication will be on node through shared memory, rather than off node through the interconnect (Figure 7.9).

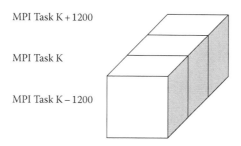

FIGURE 7.7   Communication in z dimension.

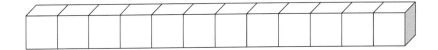

FIGURE 7.8   Default MPI task placement on 12-core node.

When designing the MPI task placement on a multicore node, it is important to minimize the surface area of the subgrid on the node. In a 3D differencing application the amount of data transferred is proportional to the surface area of the on-node grid. In the default case, the surface area of the $1 \times 1 \times 12$ subgrid, placement would be 12 up, 12 down, 12 in, 12 out, 1 right, 1 left = 50. In the $3 \times 2 \times 2$ subgrid, the surface area is 6 up, 6 down, 6 in, 6 out, 4 right, 4 left = 32. The amount of data being communicated off node is reduced by a factor of 32/50.

When we imposed the $3 \times 2 \times 2$ ordering on the MPI task layout for the node, we see a dramatic reduction in time for the larger processor runs, getting it back in line with the good scaling (Figure 7.10).

7.5 LOAD IMBALANCE

When the computation is not balanced, the cause may be poor decomposition and/or the computation performed on a node is dependent on physical conditions that are not uniform across the processors. We will first address the decomposition issues.

In a multidimensional problem, the code developer may choose to decompose the problem into planes, pencils (the points on a line containing one dimension), or as a chunk of the grid. Most codes today were

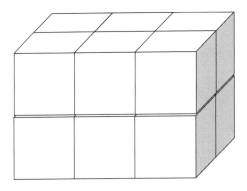

FIGURE 7.9   Preferred MPI task placement on 12 core-node.

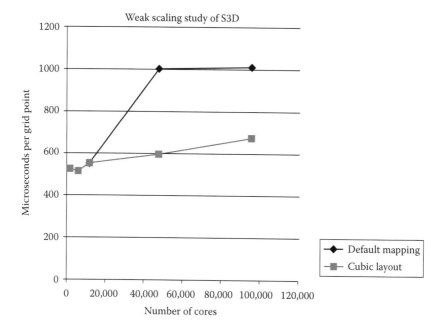

FIGURE 7.10   Comparison of original default placement versus cubic layout.

developed to run on less than 1000 processors and may only be decomposed on a plane (2D decomposition). If this is the case, the MPI parallelism will be limited to the number of planes in the grid. If there are only
1000 planes, then one cannot really consider running on 10,000 MPI
processors.

   With the existence of multicore nodes, the use of OpenMP can extend
the scalability of the 2D decomposition. If the node has 16 cores per
node, then OpenMP or Pthreads may be used on the node to achieve
scalability beyond the 1000 planes. While this may extend the scalability
somewhat, 2D decomposition is not recommended for going to 100,000
cores. Even if the nodes contain more cores, the amount of communication required when using 2D decomposition is larger than that required
in a good 3D decomposition. Consider the surface array of a plane versus
the surface area of a cube. A cube that contains the same number of
grid points as a plane has less surface area and therefore the amount of
off node communication will be less. When an application is developed
to scale to larger and larger core counts, 3D decomposition is highly
recommended.

When a code is load imbalanced, new decompositions must be considered. The best communication strategy in the world will not help a load-imbalanced application. Often, decompositions must change during the computation. A very powerful technique some developers use on massively parallel systems is adaptive mesh refinement (AMR). This technique automatically identifies load imbalance and/or severely distorted meshes and redecomposes the problem dynamically during the run. This redecomposition results in costly communication overhead and its benefit must be traded off against the benefit of attaining a more load-balanced computation.

Of course, there are much more difficult load-balancing situations. A very interesting approach has been used by CHARM++, where the work distribution is performed at runtime [13]. Of course, the use of such a system and/or technique must be considered in the initial development of the application. Retrofitting an application with CHARM++ would be very difficult. CHARM++ is a work distribution system that sends chunks of work with the necessary data to each processor. This is what is known as a master–slave approach. The system is designed to balance the work load and is extremely effective in applications such as NAMD (Not (just) Another Molecular Dynamics program).

On multicore nodes, intelligent allocation of MPI tasks to the nodes can also improve performance. Consider an ocean model mentioned previously, whose grid contains both land and ocean grid blocks. Since no computation is performed on the land grids, those will not impact memory bandwidth and/or injection bandwidth of the interconnect. By grouping some land grid blocks with ocean blocks on the same node, the ocean block benefits from having more memory bandwidth than when it shared the node with all other ocean grid blocks. The idea would be to allocate an equivalent number of ocean grids on each node. This does cause another problem in that, the nearest neighbors in the logical grid may no longer be nearest neighbors on the network. The use of this technique is very machine dependent. If the interconnect has sufficient interconnect bandwidth, this grouping of land and ocean blocks can be very beneficial.

OpenMP is used quite effectively to treat load imbalance. With OpenMP or Pthreads, the dynamic distribution of work within a node can be performed very fast, even automatically, when using the dynamic scheduling feature of OpenMP.

7.5.1 SWEEP3D

The SWEEP3D benchmark [14] is an excellent example of load imbalance. The benchmark models 1-group time-independent discrete ordinates (Sn) 3D cartesian (XYZ) geometry neutron transport problem. This is inherently load imbalanced, since the amount of computation and communication is dependent on where in the grid the processor resides. Figure 7.11 is a scaling curve run from 60 to 960 cores.

Note that we have a superlinear improvement going from 60 to 120; however, after that point, the scaling tails off. This tail off is due to load imbalance of the computation and communication. At 480 cores, we depart from the linear scaling by 24 seconds. Following is a summary profile from the CrayPat™ tool:

```
Time % | Time       |Imb. Time | Imb.    | Calls  |Experiment = 1
       |            |          | Time % |        |Group
       |            |          |         |        | Function
100.0% | 79.720128 |  --      |  --     | 4670.8 |Total
|————————————————————————————————————————————————————————————————
|58.2% | 46.403361 |  --      |  --     | 2241.4 |MPI
||———————————————————————————————————————————————————————————————
||57.2%| 45.632602 | 8.402270 | 15.6%  | 1099.2 |mpi_recv_
|| 1.0%|  0.765467 | 0.545498 | 41.7%  | 1099.2 |mpi_send_
||=============================================================
| 23.8%| 18.978924 |  --      |  --     | 2390.4 |USER
||———————————————————————————————————————————————————————————————
||23.0%| 18.307696 | 1.238828 | 6.4%   | 12.0   |sweep_
||=============================================================
| 18.0%| 14.337844 |  --      |  --     | 39.0   |MPI_SYNC
||———————————————————————————————————————————————————————————————
||10.7%|  8.534260 | 8.588305 | 50.3%  | 32.0   |mpi_allreduce_(sync)
|| 7.3%|  5.801636 | 1.133664 | 16.4%  | 4.0    |mpi_bcast_(sync)
```

Note that MPI_SYNC is showing 14 seconds of load imbalance. Most of that come from the MPI sends/receives and a small amount from the sweep routine where most of the computation is performed. One potential improvement to this load imbalance is to place the tasks that perform a lot of communication with the tasks that do not perform much communication on the same node. In this case, each node has eight cores. This will even out the need for limited resources required on each node.

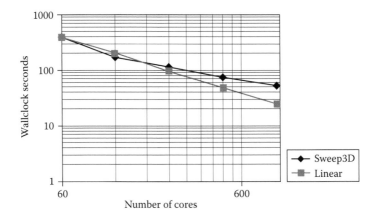

FIGURE 7.11   Scaling of SWEEP3D.

Figure 7.12 gives a little insight into the issues of scaling SWEEP3D.

In this case, the load imbalance is included in the MPI_SYNC plus the MPI-send-recv. Since this is a nonblocking receive, the processor time waiting in the receive is attributed to the MPI_RECV. At the largest processor count, the load imbalance accounts for 71% of the time.

By running a larger problem, the amount of the computation would increase, and we would see that the load imbalance does not impact the scaling until it becomes the dominant portion of the total time.

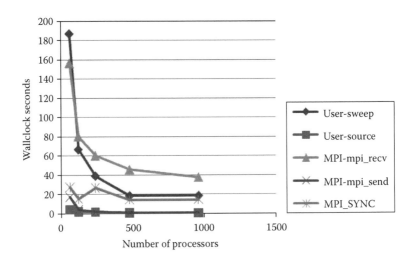

FIGURE 7.12   Scaling of individual components of SWEEP3D.

Once the load balance is fixed, new statistics should be obtained to see if the bottlenecks have changed. If computation is now the bottleneck, processor and node performance needs to be considered. If communication is still the bottleneck, then the individual message-passing functions must be examined. Always deal with the load imbalance before addressing any other bottleneck in the system. Whenever changes are made, make sure that the answers are correct and the scaling runs are redone.

7.6 COMMUNICATION BOTTLENECKS

7.6.1 Collectives

When a majority of the communication within an application is used in a collective, MPI_ALLREDUCE, MPI_REDUCE, MPI_ALLTOALL, MPI_VGATHER, MPI_VSCATTER, and so on, the use of OpenMP and/or Pthreads on the node should be considered. Collectives take a larger and larger fraction of time as the number of MPI tasks is increased. By employing OpenMP on the node, the number of MPI tasks can be significantly reduced, thus enabling scaling to a higher number of cores. If the number of MPI tasks is reduced to one per socket, then OpenMP must be incorporated into the application to achieve parallelism within the socket.

Another powerful technique is combining collectives whenever possible. Oftentimes, numerous global reduces are performed at each iteration of time. These reduces should be combined into a single call to MPI_REDUCE, increasing the number of elements to be reduced accordingly. This approach requires that the elements to be reduced are packed into a contiguous section of memory. The time to perform an MPI_REDUCE or MPI_ALLREDUCE for 100 variables is only slightly longer than the time to perform the operation for one variable.

7.6.1.1 Writing One's Own Collectives to Achieve Overlap with Computation

Perhaps the biggest inhibitor to scaling to very large processor counts is MPI_ALLTOALL of large data sets. Many CFD codes must perform a 3D fast Fourier transform (FFT) that requires FFTs on X, Y, and Z. Depending on the decomposition of the data, two or three transposes of the data must be performed. As the problem size becomes larger, these transposes become larger and soon the global bandwidth of the interconnect becomes a major bottleneck. Typically, the transposes are performed as a *blocked* operation—that is, all other operations stop until the transpose is finished. A solution to this dilemma is to replace the MPI_ALLTOALL with sends/receives performed in parallel with the computation for the FFT. This rewrite is difficult using MPI; however, the ability to put data into another processors memory using co-arrays and/or Unified Parallel C (UPC) make the task easier.

7.6.2 Point to Point

When a majority of the time is spent in a point-to-point communication, the developer must examine the cause of that increased time. This is when the message-passing statistics are needed to identify what messages are taking most of the time. Is the bottleneck due to a lot of small messages? Is the bottleneck due to large messages? Can the communication be overlapped with computation? Can the communication be overlapped with other communication? If the size of the message that takes most of the time is very small, then the latency of the interconnect is the bottleneck and perhaps all the small messages that go to a given MPI task can be combined into a single, larger message. If the size of the messages that takes most of the time is large, then the interconnect bandwidth could be the bottleneck and perhaps the messages can be broken up into smaller messages and sent asynchronously overlapped with computation.

7.6.3 Using the "Hybrid Approach" Efficiently

The "hybrid approach" consists of combining OpenMP with MPI. In Chapter 8, numerous cases are given where OpenMP was used to scale an application on the node. Now, the question is when can OpenMP be introduced to an existing MPI program to improve its scalability? This issue is more difficult to address, since introducing OpenMP takes processors away from MPI. If an MPI application obtains a factor of 2.5 in performance in going from 1024 to 4096 cores, the intent would be to take the 1024-core MPI run, assign a single MPI task to each quad core socket, and then run OpenMP across the cores within the socket. We know from Amdahl's law that attaining more than a factor of 2.5 out of a parallelism of 4 would require that 80–90% of the application must be parallelized. The incorporation of OpenMP into an MPI application is not that easy and represents a significant undertaking. However, looking forward, as chips contain more and more shared processors and memory and interconnect bottlenecks continue to plague the MPI performance, some sort of threading on the node is desirable.

The major areas where OpenMP can potentially help MPI applications are the following:

1. The biggest reason would be to obtain higher performance for a strong scaling application that has run out of scaling due to the decomposition of the problem

2. OpenMP can be used to improve load imbalance

3. OpenMP can reduce network traffic

4. OpenMP can reduce memory requirements

5. OpenMP can reduce memory bandwidth requirements

Achieving good scaling with OpenMP on a shared memory parallel system is very difficult in most cases. The first and foremost problem is that the programmer must parallelize the code which uses an appreciable amount of the compute time.

7.6.3.1 SWIM Benchmark

In the previous chapter, the SWIM benchmark was examined. This example gives us an easy test of all-MPI versus the hybrid approach. Figure 7.13 is the all-MPI scaling curve.

We now apply OpenMP to the three major loops and obtain the following performance for 2-way OpenMP and 4-way OpenMP (Figure 7.14).

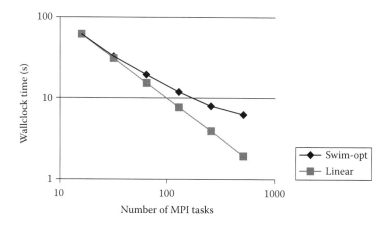

FIGURE 7.13   Scaling of optimized SWIM benchmark.

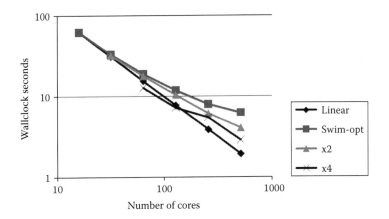

FIGURE 7.14   Scaling of optimized SWIM benchmark using OpenMP on 2 and 4 threads.

This is an ideal situation for hybrid computation. The OpenMP is improving the scaling by reducing the number of MPI tasks. While the performance is still not linear, OpenMP has given us a factor of two at 512 cores.

Many comparisons have been made between all-MPI and the hybrid approach that show that all-MPI is the best approach at low processor counts. This is the wrong approach, the comparisons should be made in the area where all-MPI scaling stops. For example, in the following QCD code, BQCD, we see that higher levels of OpenMP threading do not perform well at low processor counts; however, it is superior as the processor count grows. The idea is to extend the scalability of the MPI code with OpenMP (Figure 7.15).

7.7 OPTIMIZATION OF INPUT AND OUTPUT (I/O)

Would it not be great if the results of a very large parallel application were simply "yes" or "no?" Unfortunately, that is not the case and even more important these days is the ability to generate an animation of a computation to better visualize the phenomenology behind the computation. Additionally, many of the large applications that are run across a large parallel system must be written to produce restart files. If a particular problem is going to take 200 hours of compute time, the likelihood of a system staying up for more than 8 days is small; hence, the application should write checkpoints or restart files that can be used to restart the computation from and continue running. So a 200-hour computation could be divided into 20 10-hour runs. All the data required to obtain the

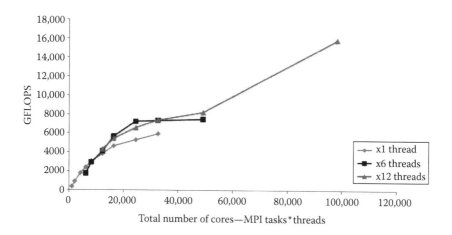

FIGURE 7.15   BQCD hybrid performance.

input for a movie and/or a restart file means a lot of output for an application. Some applications may be large enough that its input files are enormous and the reading of the data may take some significant amount of time. So as we run increasingly larger computations that require and generate increasingly larger data sets, we start seeing I/O problems.

7.7.1   Parallel File Systems

There are many parallel file systems being used on large MPP systems today. The most prevalent seems to be Lustre [19]. This file system is built upon the notion that there are multiple nodes (object storage servers (OSSs)) in the MPP that handle the transfer of the data between the user's application to the disks. The actual file operations are performed by the meta-data server (MDS) which opens, positions, and closes the files. On very large systems, such as the Cray XT™ systems there could be hundreds of OSSs available for transfer data. When an application employs multiple writers as mentioned above, there is a potential to have parallel writes or reads going across the network through the OSSs out to the disks.

A serious limitation of Lustre is the existence of a single MDS. This processor, or multicore node, is used by all of the users on the MPP to open, close, and position the file pointers. The MDS is also used when users issue any Linux® command that queries a file. On a very large MPP system using Lustre, if an application performs inefficient I/O, such as, having 100,000 processors opening the same file, other users experience slow response to a simple *ls–ltr* Linux command. Writing inefficient I/O is similar to being a polluter of our environment: your bad practices end up impacting the lives of everyone living around you.

7.7.2 An Inefficient Way of Doing I/O on a Large Parallel System

Today, most applications that have been running on hundreds of processors typically read data on a single file and use MPI_BCAST to send the input data to the other processors. Additionally, when output is required, all the processors may send its data to MPI task 0, and have the single processor output the data. As the application is scaled up to increasingly larger processor counts, this is not a scalable solution. All the processors not involved in the I/O have to wait until the single processor doing the I/O completes. Compared to computation, I/O is a very costly process and the amount of time to send the data to and from the processor doing the I/O can be significant. The I/O bandwidth is also limited by the bandwidth from a single processor to the disk. If an application runs on a parallel system, it should take advantage of parallel I/O.

Many application developers may go to the other extreme, where every processor reads the same input data and writes its data into separate data files (or direct access files). Actually this approach works reasonably well up to 5000–6000 processors. There are two issues with this approach. First, when 5000–6000 processors access the same data file for input or output to a direct access file, the metadata server becomes extremely busy handling all the individual requests. Unfortunately, the overloading of the metadata server on a large parallel system would also impact other users on the system who access the shared parallel file. The second issue is the shear number of files that get generated if every processor writes a distinct file. Managing the disparate files after the application terminates can be a hassle. When applications start scaling beyond this range a new approach must be considered.

7.7.3 An Efficient Way of Doing I/O on a Large Parallel System

Consider the following approach. Designate an I/O processor for a group of processors. A good selection mechanism is to take the square root of the number of MPI tasks and use that in many I/O processors. So for 100,000 MPI tasks, there would be 1000 I/O groups with a single I/O processor in each group. In this way, the application can take advantage of a parallel file system without overloading it with 100,000 separate requests. Within each I/O group there are only 1000 processors, and data coalescing can be performed in parallel across the groups. I/O can also be done in parallel.

This is a compromise between two inefficient I/O approaches which are prevalent in the parallel processing community.

EXERCISES

1. Several methods exist for performing a global summation of a single variable. What is the combining tree approach? On an application using 65,536 processors across 8192 nodes with two quad-core sockets, how many levels of the tree are on nodes and how many are between nodes?

2. What is superlinear scaling? Why does it occur?

3. Why is grid partitioning usually not a performance problem? When can it be a problem?

4. Why is load imbalance often the cause of a large amount of time spent in reduction operations?

5. How might MPI process location affect performance of a 3D code with nearest-neighbor communication?

6. On a strong scaling parallel run, the point where the working set of an application fits into cache exhibits a superlinear speedup. Can this be achieved with an OpenMP? What are the requirements on an OpenMP loop that might exhibit such a superlinear speedup.

7. How can load imbalance be addressed with an OpenMP?

8. How can an OpenMP improve the scaling of an all-MPI application?

9. Many applications cannot always be written to access the innermost dimension in the inner DO loop. For example, consider an application that performs FFTs across the vertical lines of a grid and then performs the transform across the horizontal lines of a grid. In the FFT example, code developers may use a transpose to reorder the data for better performance when doing the FFT. The extra work introduced by the transform is only worth the effort if the FFT is large enough to benefit from the ability to work on local contiguous data. Using a library FFT call that allows strides in the data, measure the performance of the two approaches (a) not using a transpose and (b) using a transpose and determine the size of the FFT where (b) beats (a).

10. What are the advantages and disadvantages of using a single node to perform all I/O? How is this different when using hundreds of nodes versus thousands of nodes?

11. What are the advantages and disadvantages of having each node perform its own I/O?

CHAPTER 8

Node Performance

I N THIS CHAPTER, WE look at the SPEComp® benchmark suite [14]. These tests give us a wide variety of examples, some run very well and others do not run so well. In each of the cases, we will examine the code and determine the reasons for the obtained performance. The hope is that these codes would give the reader an idea of the issues involved when using OpenMP™ on the node. The effective use of OpenMP on the node is a precursor to obtaining good Hybrid code, which was discussed in Chapter 7.

8.1 WUPPERTAL WILSON FERMION SOLVER: WUPWISE

The first example we will examine is the WUPWISE code. "WUPWISE" is an acronym for "Wuppertal Wilson Fermion Solver," a program in the area of lattice gauge theory (quantum chromodynamics).

The results for this application show excellent scaling (Figure 8.1).

In this plot, we show the routines that make up over 90% of the execution time. The overall runtime achieves a factor of 3.55 in going from 2 to 8 threads. Why does this application get such excellent improvement? There are several reasons.

1. A large percent of the original code is parallelized; 99.9% of the original execution time is now being threaded.

2. Most of the parallel regions have very large granularity. Following is a snippet of one of the parallel regions. Notice the calls within the

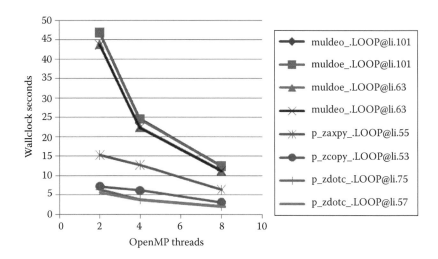

FIGURE 8.1   Scaling of major routines in WUPWISE.

parallel DO. This is the best way to get superior performance. Utilize the OpenMP at the highest level of the call chain.

```
C$OMP PARALLEL
C$OMP+    PRIVATE(AUX1, AUX2, AUX3),
C$OMP+    PRIVATE(I, IM, IP, J, JM, JP, K, KM, KP, L, LM, LP),
C$OMP+    SHARED(N1, N2, N3, N4, RESULT, U, X)
C$OMP DO
        DO 100 JKL = 0, N2 * N3 * N4 - 1
        L = MOD(JKL/(N2 * N3), N4) + 1
        LP = MOD(L,N4) + 1
         K = MOD(JKL / N2, N3) + 1
         KP = MOD(K,N3) + 1
          J = MOD (JKL, N2) + 1
          JP = MOD(J,N2) + 1
          DO 100 I = (MOD(J + K + L + 1,2) + 1),N1,2
           IP = MOD(I,N1) + 1
           CALL GAMMUL(1,0,X(1,(IP + 1)/2,J,K,L),AUX1)
           CALL SU3MUL(U(1,1,1,I,J,K,L),'N',AUX1,AUX3)
           CALL GAMMUL(2,0,X(1,(I + 1)/2,JP,K,L),AUX1)
           CALL SU3MUL(U(1,1,2,I,J,K,L),'N',AUX1,AUX2)
           CALL ZAXPY(12,ONE,AUX2,1,AUX3,1)
           CALL GAMMUL(3,0,X(1,(I + 1)/2,J,KP,L),AUX1)
           CALL SU3MUL(U(1,1,3,I,J,K,L),'N',AUX1,AUX2)
```

```
        CALL ZAXPY(12,ONE,AUX2,1,AUX3,1)
        CALL GAMMUL(4,0,X(1,(I+1)/2,J,K,LP),AUX1)
        CALL SU3MUL(U(1,1,4,I,J,K,L),'N',AUX1,AUX2)
        CALL ZAXPY(12,ONE,AUX2,1,AUX3,1)
        CALL ZCOPY(12,AUX3,1,RESULT(1,(I+1)/2,J,K,L),1)
100 CONTINUE
C$OMP END DO
```

3. Some of the routines being parallelized are somewhat memory bound. As we increased the number of threads on the node, we observed a larger impact on the memory bandwidth required to supply the operands to the processor. In the case where the code is doing Level 2 basic linear algebra subroutines (BLAS) which stress the memory bandwidth, the performance degrades. Note the following table that gives speedup over 2 threads for the various OpenMP regions:

	4 Threads	8 Threads
muldoe_.LOOP@li.101	1.95	3.88
muldoe_.LOOP@li.101	1.94	3.81
muldoe_.LOOP@li.63	1.96	3.90
muldeo_.LOOP@li.63	1.96	3.90
p_zaxpy_.LOOP@li.55	1.20	2.39
p_zcopy_.LOOP@li.53	1.16	2.36
p_zdotc_.LOOP@li.75	1.63	3.10
p_zdotc_.LOOP@li.57	1.52	2.93

Note the last four regions that are Level 2 BLAS are only getting 2–3 in improvement going from 2 to 8 threads. This is due to the fact that the Level 2 BLAS are very memory intensive. The first four regions are actually Level 3 BLAS which are more dependent on matrix multiply, which is not as dependent upon memory bandwidth.

8.2 SWIM

SWIM is an ancient benchmark that solves a finite-difference representation of the shallow water equations. As we can see from Figure 8.2, the performance of this application is not good.

In this application, we only obtain an overall improvement of 3 going from 1 to 8 processors. It is also interesting that we obtain a factor of 2

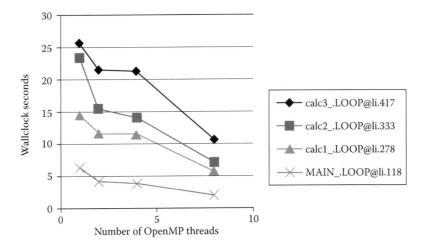

FIGURE 8.2   Scaling of major routines in SWIM.

going from 4 to 8 processors. There are good reasons for the observed performance.

1. As in WUPWISE, 99.9% of the original code is being parallelized.

2. The granularity of the 3–4 loops is large enough to amortize the overhead of parallelization.

3. The real problem with this is the memory bandwidth demands of the parallelized loops. Consider the following sample loop from SWIM:

```
!$OMP PARALLEL DO
      DO 100 J = 1,N
      DO 100 I = 1,M
      CU(I+1,J) = .5D0*(P(I+1,J) + P(I,J))*U(I+1,J)
      CV(I,J+1) = .5D0*(P(I,J+1) + P(I,J))*V(I,J+1)
      Z(I+1,J+1) = (FSDX*(V(I+1,J+1) -V(I,J+1)) -FSDY*(U(I+1,J+1)
     1           -U(I+1,J)))/(P(I,J) + P(I+1,J) + P(I+1,J+1) + P(I,J+1))
      H(I,J) = P(I,J) + .25D0*(U(I+1,J)*U(I+1,J) + U(I,J)*U(I,J)
     1              +V(I,J+1)*V(I,J+1) + V(I,J)*V(I,J))
  100 CONTINUE
```

This is one of the important kernels in SWIM. This example is extremely memory bandwidth limited. When we look at the Level 1 cache hit ratio calc2_OpenMP loop compared with the speedup on 4 and 8 for the threads, we see an interesting correlation. The following table gives speedups over a single thread. As the hit ratio goes up, the speedup we obtain from the OpenMP goes up.

DO Loop	Level 1 Hit Ratio	4	8
100	91.00%	1.209954	2.434201
200	93.40%	1.662905	3.308106
300	90.30%	1.268123	2.529215

Remember from Chapter 2: A cache line contains eight 64-bit operands. If we use every element of the cache line, our cache hit ratio will be 7/8 or 87.5%, since we will have a miss every 8 fetches. The hit ratio in the table shows that we obtain little cache reuse, since we achieve only hit ratios in the low 90s. Since we obtain poor cache utilization, we demand more memory bandwidth and as the number of threads increase, this application quickly becomes memory limited. Confirmation of this conclusion is the increase in performance going from 4 to 8 cores. Note in Figure 8.3, of a two-quad-core Opteron™ system, that 1, 2, and 4 cores run on a single socket of the node.

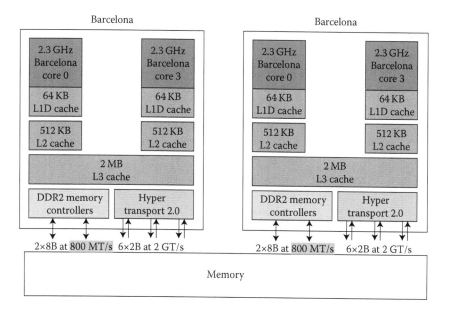

FIGURE 8.3　Two-quad-core Opteron system design.

As we increase the performance to 8 cores, the bandwidth is doubled. On the 4 core run, the memory bandwidth through a single socket was being used, as we went to 8 cores, we ran across two sockets and doubled the effective memory bandwidth, and thus result in a factor of two in performance.

Observe that the performance of the 4-thread run did not improve over the 2 threads and when we go to 8 threads we obtain a good improvement. The reason for this is that when we ran with the 4 threads, all four threads ran on one of the two sockets, while the 8-thread run ran on both sockets and it used both memory paths. We can make the 4-thread run use both sockets memory by using an option on the run command:

aprun − n 1 − d 4 − cc 0,1,4,5 ./a.out

that specifies which socket to run the thread on. In this case, threads 0 and 1 run on cores 0 and 1 which are in the first socket and threads 3 and 4 run on cores 4 and 5 which are on the second socket. For example on the Cray XT™ platform, the run command runs 2 threads on the first socket (0,1) and 2 on the second socket (4,5). Using this for the 4-thread run would give better timings; however, when we go to 8 threads, we would once again be memory limited.

8.3 MGRID

MGRID is one of the NASA parallel benchmarks, and it applies a multigrid solver on a three-dimensional grid. From the graph, we once again see the performance illustrative of a memory-bound application. When we apply the cache block strategy discussed in Chapter 5, we obtain a significant improvement on the most memory bound 4-thread case and a lesser improvement at 2 and 8 threads (Figures 8.4 and 8.5).

The important lesson is that poor cache utilization will steal OpenMP scalability, simply because it increases the reliance on memory bandwidth which is the rarest of all commodities on the node.

8.4 APPLU

APPLU is another NASA parallel benchmark which performs the solution of five coupled nonlinear PDEs on a three-dimensional logically structured grid, using an implicit psuedotime marching scheme, based on a two-factor approximate factorization of the sparse Jacobian matrix. Figure 8.6 shows that some routines scale very well while others do not. The overall performance is 3.91 on 8 threads. The major computational OpenMP region only achieves 2.75 improvement in going from 1 to 8 threads. While the parallel

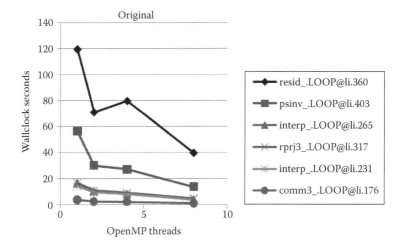

FIGURE 8.4   Scaling of major routines in MGRID.

region is at a very high level, there is a !$OMP BARRIER in the middle of the major high-level loops within the parallel regions.

```
!$omp parallel
!$omp& default (shared)
!$omp& private (i0, i1, ipx, ipy, j0, j1, k, l, mt, nt, npx, npy)
!$omp& shared (nx, ny, nz, omega)
        O       o       o
```

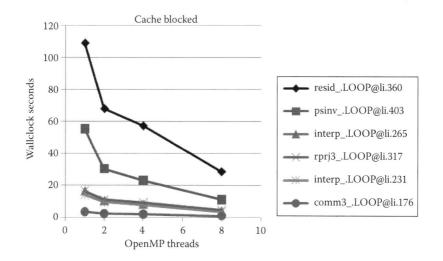

FIGURE 8.5   Scaling of cache block routines in MGRID.

```
      DO l = 2,  npx + npy + nz - 3
      k = l - ipx - ipy
      if ((1 .lt. k) .and. (k .lt. nz)) then
c---------------------------------------------------------------
c    form the lower triangular part of the jacobian matrix
c---------------------------------------------------------------
            call jacld (i0, i1, j0, j1, k)


c---------------------------------------------------------------
c    perform the lower triangular solution
c---------------------------------------------------------------
            call blts(isiz1, isiz2, isiz3,
   >                        nx, ny, nz, i0, i1, j0, j1, k,
   >                        omega,
   >                        rsd, tv,
   >                        a, b, c, d)
      end if
!$omp barrier
      end do
      o    o    o
!$omp end parallel
```

   This causes the threads to synchronize within the parallel execution of the region. This synchronization introduces overheads and effectively reduces the granularity of the parallel region. The barriers are required to assure accurate update of the matrices in the routines that are called after the barrier. For example, if the array $A(i,j,k)$ is updated in the DO loop prior to the barrier and then in the loop after the barrier, another array is updated using $a(i,j,k-1)$ and/or $a(i,j,k+1)$; then there is data dependency between the two loops. To remove this data dependency, the user must assure that all threads finish the first loop before proceeding to the second loop. Unfortunately, in this case, the barrier causes a degradation of performance. With this major region achieving less than 50% of the desired performance increase, the overall parallel performance of this application suffers (Figure 8.6).

8.5 GALGEL

The GALGEL application is a particular case of the GAMM (Gesellschaft fuer Angewandte Mathematik und Mechanik) benchmark devoted to the numerical analysis of oscillatory instability of convection in

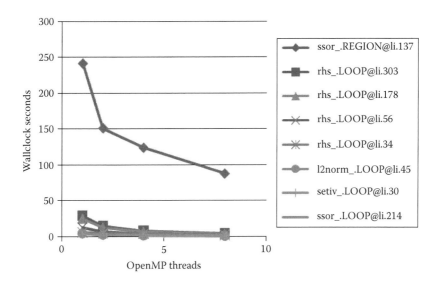

FIGURE 8.6   Scaling of major routines in APPLU.

low-Prandtl-number fluids [1]. Once again, we have a variation on the speedup of the individual routines and the overall speedup of 3.55 out of 8.

From Figure 8.7, we see that the main culprit is the routine main. On a single thread the main takes only 20 seconds and on 8 threads it takes 50 seconds. In the examination of main we see that there is a lot of data initialization and OpenMP SECTION directives, which give each thread a section of computation to perform.

```
! ***** Begin forming the Galerkin system *****************

! ..... Evaluation of inner products of polynomials ......

!$OMP PARALLEL SECTIONS
!$OMP SECTION
        C a l l    PolLin (Nx, Ny, NKx, NKy)
!$OMP SECTION
        C a l l    PolNel (Nx, Ny, NKx, NKy)

! .... Introducing of the boundary conditions ...........

!$OMP SECTION
        C a l l    FunNS (Nx, Ny)
!$OMP SECTION
        C a l l    FunHT (NKx, NKy)
!$OMP END PARALLEL SECTIONS
```

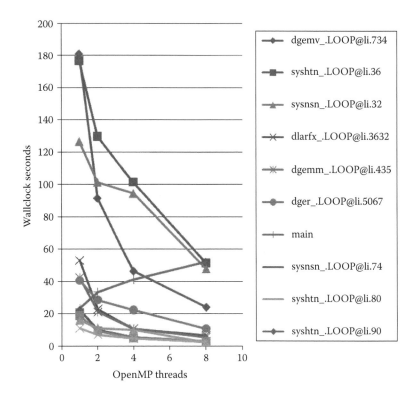

FIGURE 8.7   Scaling of major routines in GALGEL.

The first issue is that this OpenMP construct only uses 4 threads, but, more importantly, these subroutines are very small, and the overhead of running these four calls in parallel, which increases as we increase our threads, is too large to get any benefit out of parallelizing this structure. It would be far better to remove these OpenMP directives and just perform these calls on the master thread (Figure 8.7).

8.6 APSI

APSI is a mesoscale hydrodynamic model. Its original performance is not as good as it should be and the inefficiency is due to a relatively unknown issue when using OpenMP. Figure 8.8 illustrates the performance attained from the original run.

It is interesting that the performance, when going from 1 to 2 threads, is actually very good; however, above 2 threads the performance is horrible. This is an excellent example of "false sharing," that is, when more than one thread accesses the same cache line. When this happens, the ownership of

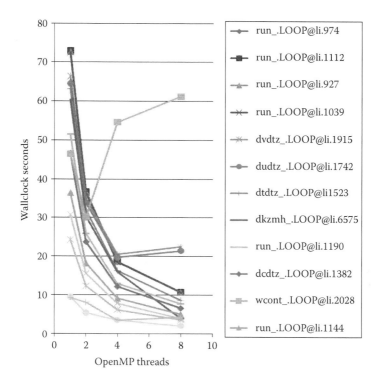

FIGURE 8.8   Scaling of major routines in APSI.

the cache line, which is required to write into the cache line is in dispute and this simple case kills the performance of this application (Figure 8.8).

Note in the graph that wcont_.LOOP@li.2028 actually grows in time as the number of cores increases. In examining this looping structure, we see a use of a shared array to hold information to be summed outside the DO loop. Arrays WWIND1 and WSQ1 are dimensioned by the number of threads and each OpenMP thread updates its part of the array. Each thread runs on a different core and each time the array is updated, the cache line containing that array must be fetched to its Level 1 cache. Since there can only be one cache line in each of the cores Level 1 cache containing the array, there is significant thrashing between accessing of the arrays in question. The solution which is very simple is to dimension each array large enough that each core accesses a different cache line. For example, if we dimension arrays WWIND1(32,NUMTHREADS), WSQ1(32,NUMTHREADS) and access the second dimension in the loop such as WWIND1(1,I) and WSQ1(1,I), each reference will be on a different cache line.

```
C
      DO 25 I = 1, NUMTHREADS
         WWIND1(I) = 0.0
         WSQ1(I) = 0.0
  25     CONTINUE


!$OMP PARALLEL
!$OMP + PRIVATE(I,K,DV,TOPOW,HELPA1,HELP1,AN1,BN1,CN1,MY_CPU_ID)
        MY_CPU_ID = OMP_GET_THREAD_NUM() + 1
!$OMP DO
      DO 30 J = 1, NY
         DO 40 I = 1, NX
            HELP1(1) = 0.0D0
            HELP1(NZ) = 0.0D0
            DO 10 K = 2, NZTOP
               IF(NY.EQ.1) THEN
                  DV = 0.0D0
                          ELSE
                  DV = DVDY(I,J,K)
               ENDIF
               HELP1(K) = FILZ(K)*(DUDX(I,J,K) + DV)
  10        CONTINUE
C
C    SOLVE IMPLICITLY FOR THE W FOR EACH VERTICAL LAYER
C
            CALL DWDZ(NZ,ZET,HVAR,HELP1,HELPA1,AN1,BN1,CN1,ITY)
            DO 20 K = 2, NZTOP
               TOPOW = UX(I,J,K)*EX(I,J) + VY(I,J,K)*EY(I,J)
               WZ(I,J,K) = HELP1(K) + TOPOW
               WWIND1(MY_CPU_ID) = WWIND1(MY_CPU_ID) + WZ(I,J,K)
               WSQ1(MY_CPU_ID) = WSQ1(MY_CPU_ID) + WZ(I,J,K)**2
  20        CONTINUE
  40     CONTINUE
  30  CONTINUE
!$OMP END DO
!$OMP END PARALLEL


      DO 35 I = 1, NUMTHREADS
         WWIND = WWIND + WWIND1(I)
         WSQ = WSQ + WSQ1(I)
  35     CONTINUE
```

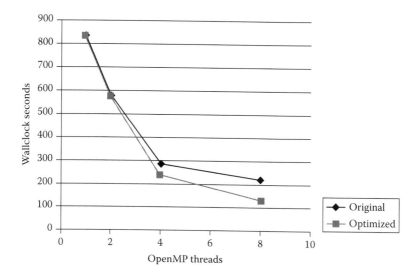

FIGURE 8.9   Comparison of original and restructure scaling of APSI.

This change was also made in the dvdtz_LOOP@li.1915 and dudtz_LOOP@li.1742. The results are quite remarkable for such a simple change (Figure 8.9):

	Original	Optimized
1	830	830
2	582	574
4	284	234
8	220	137

The scaling of this example has been improved from 3.77 out of 8 to 6.05 out of 8.

8.7 EQUAKE

EQUAKE is a program that simulates the propagation of elastic waves in large, high heterogeneous valleys, such as California's San Fernando Valley, or the Greater Los Angeles Basin. This is a C program and exhibits poor OpenMP scaling of 4.45 going from 1 to 8 threads (Figure 8.10).

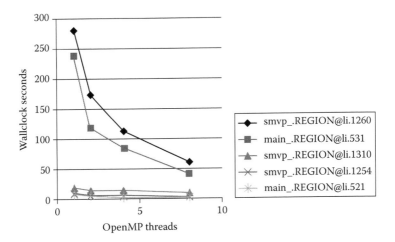

FIGURE 8.10   Scaling of major routines in EQUAKE.

Notice the following code which is the OpenMP region 521:

```
#pragma omp for
    for (i = 0; i < ARCHnodes; i ++) {
        c0[0] = ARCHcoord[i][0];
        c0[1] = ARCHcoord[i][1];
        c0[2] = ARCHcoord[i][2];
        d1 = distance(c0, Src.xyz);
        d2 = distance(c0, Src.epixyz);

        if (d1 < bigdist1[my_cpu_id]) {
            bigdist1[my_cpu_id] = d1;
            temp1[my_cpu_id] = i;
        }

        if (d2 < bigdist2[my_cpu_id]) {
            bigdist2[my_cpu_id] = d2;
            temp2[my_cpu_id] = i;
        }
    }
```

Once again, we have all the threads fighting over the cache line that contains bigdist1 and bigdist2, the same issue that we addressed for EQUAKE. The second OpenMP region in the main has a significant issue with load imbalance. Note that a significant amount of the computation appears to perform only under certain circumstances. In this case, it may be wise to try some kind of dynamic scheduling of the threads executing this loop.

```
    if (Src.sourcenode != 0) {
#pragma omp parallel for private(cor,j,k,xc,vertices)
    for (i = 0; i < ARCHelems; i++) {
      for (j = 0; j < 4; j++)
        cor[j] = ARCHvertex[i][j];

      if (cor[0] == Src.sourcenode || cor[1] == Src.sourcenode ||
         cor[2] == Src.sourcenode || cor[3] == Src.sourcenode) {

      for (j = 0; j < 4; j++)
        for (k = 0; k < 3; k++)
          vertices[j][k] = ARCHcoord[cor[j]][k];

      centroid(vertices, xc);

      source_elms[i] = 2;
      if (point2fault(xc) >= 0)
      source_elms[i] = 3;

      }
    }
  }
```

8.8 FMA-3D

FMA-3D is a finite-element method computer program designed to simulate the inelastic, transient dynamic response of three-dimensional solids. The performance is reasonably good, obtaining a factor of 5.31 going from 1 to 8 threads. The following chart shows the scaling of the individual routines. This is an excellent implementation of OpenMP at a very high level and the major computing achieves an excellent 7.09 out of 8 (Figure 8.11).

The OMP regions that are not scaling have very bad memory access issues which destroys the OpenMP scaling. Consider the following DO loop:

```
!$OMP PARALLEL DO DEFAULT(SHARED) PRIVATE(N)
      DO N = 1,NUMRT
        MOTION(N)%Ax = NODE(N)%Minv * FORCE(N)%Xext-FORCE(N)%Xint)
        MOTION(N)%Ay = NODE(N)%Minv * FORCE(N)%Yext-FORCE(N)%Yint)
        MOTION(N)%Az = NODE(N)%Minv * FORCE(N)%Zext-FORCE(N)%Zint)
      ENDDO
!$OMP END PARALLEL DO
```

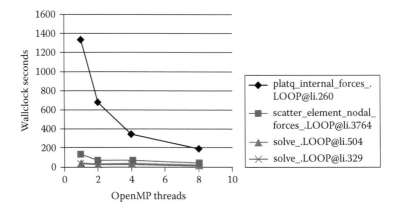

FIGURE 8.11   Scaling of major routines in FMA3D.

Following is a summary of hardware counters for this loop:

```
==========================================================================
USER/solve_.LOOP@li.329
--------------------------------------------------------------------------

Time%                                      4.5%
Time                                       12.197115 secs
Imb.Time                                   0.092292 secs
Imb.Time%                                  1.0%
Calls                      42.9 /sec       523.0 calls
PAPI_L1_DCM                13.700M/sec     167144470 misses
PAPI_TLB_DM                0.448M/sec      5460907 misses
PAPI_L1_DCA                89.596M/sec     1093124368 refs
PAPI_FP_OPS                52.777M/sec     643917600 ops
User time (approx)         12.201 secs     32941756956 cycles  100.0%Time
Average Time per Call                      0.023321 sec
CrayPat Overhead : Time     0.0%
HW FP Ops / User time      52.777M/sec     643917600 ops   0.5%peak(DP)
HW FP Ops/WCT              52.777M/sec
Computational intensity    0.02 ops/cycle  0.59 ops/ref
MFLOPS (aggregate)         52.78M/sec
TLB utilization            200.17 refs/miss  0.391 avg uses
D1 cache hit,miss ratios   84.7% hits      15.3% misses
D1 cache utilization (M)   6.54 refs/miss  0.817 avg uses

==========================================================================
```

Note the poor TLB utilization: any TLB reference/miss below 512 is suspect. While the DO loop appears as if it is contiguous in memory, since we are accessing elements of a derived type, there is a stride of the number of elements within the derived type.

```
TYPE :: motion_type
   REAL(KIND(0D0))  Px    ! Initial x-position
   REAL(KIND(0D0))  Py    ! Initial y-position
   REAL(KIND(0D0))  Pz    ! Initial z-position
   REAL(KIND(0D0))  Ux    ! X displacement
   REAL(KIND(0D0))  Uy    ! Y displacement
   REAL(KIND(0D0))  Uz    ! Z displacement
   REAL(KIND(0D0))  Vx    ! X velocity
   REAL(KIND(0D0))  Vy    ! Y velocity
   REAL(KIND(0D0))  Vz    ! Z velocity
   REAL(KIND(0D0))  Ax    ! X acceleration
   REAL(KIND(0D0))  Ay    ! Y acceleration
   REAL(KIND(0D0))  Az    ! Z acceleration
END TYPE
TYPE (motion_type), DIMENSION(:), ALLOCATABLE :: MOTION
```

Rather than the arrays being dimensioned within the derived type, the derived type is dimensioned. This results in each of the arrays having a stride of 12, which hurts both TLB and cache utilization.

8.9 ART

The performance of ART is extremely good, obtaining a factor close to 7 out of 8 threads. The granularity is very large and the computations are very cache-friendly (Figure 8.12).

Compare the following hardware counters obtained on this benchmark to the previous FD3D example.

```
========================================================================
USER/scan_recognize.REGION@li.1545
------------------------------------------------------------------------
Time%                                        99.0%
Time                                         113.561225 secs
Imb.Time                                     0.000254 secs
Imb.Time%                                    0.0%
Calls               0.0 /sec                 1.0 calls
PAPI_L1_DCM         1.776M/sec               201666688 misses
PAPI_TLB_DM         0.518M/sec               58810944 misses
PAPI_L1_DCA         1675.833M/sec            190312617117 refs
PAPI_FP_OPS         372.458M/sec             42297398657 ops
```

```
User time (approx)       113.563 secs       306620148997 cycles  100.0%Time
Average Time per Call                        113.561225 sec
CrayPat Overhead : Time  0.0%
HW FP Ops / User time    372.458M/sec        42297398657 ops       3.4%peak(DP)
HW FP Ops / WCT          372.458M/sec
Computational intensity  0.14 ops/cycle      0.22 ops/ref
MFLOPS (aggregate)       372.46M/sec
TLB utilization          3236.01 refs/miss   6.320 avg uses
D1 cache hit,miss ratios 99.9% hits          0.1% misses
D1 cache utilization (M) 943.70 refs/miss    117.962 avg uses

========================================================================
```

TLB and cache usage is outstanding which results in less memory traffic, and thus as the number of threads increase, the scaling is very good.

8.10 ANOTHER MOLECULAR MECHANICS PROGRAM (AMMP)

AMMP stands for "another molecular mechanics program." The application scales reasonably well obtaining a factor of 5.5 out of 8 threads. The issue that prevents better scaling is load imbalance caused by a critical region. In this case, the critical region is required to allow the various threads to add their contribution into shared variables, required for the

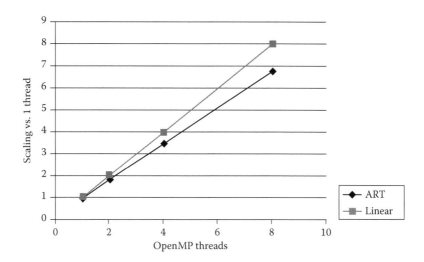

FIGURE 8.12   Scaling of ART from 1 to 8 threads.

correctness of the algorithm. With the assistance of a sampling tool, we find the following information:

```
Samp %       |    Samp   | Imb. |  Imb. | Group
             |           | Samp | Samp %| Function
             |           |      |       | Line
             |           |      |       | PE.Thread = 'HIDE'
    100.0%   |   550261  |  --  |   --  | Total
|-------------------------------------------------------------
|    77.1%   |   424368  |  --  |   --  |USER
| |-----------------------------------------------------------
| |  76.3%   |   419719  |  --  |   --  |mm_fv_update_nonbon
| | |---------------------------------------------------------
3| |   2.7%  |    14776  | 4475.88 | 26.6% |line.684
3| |   3.5%  |    19180  |  746.00 |  4.3% |line.688
3| |   8.7%  |    47677  | 2647.50 |  6.0% |line.690
3| |   1.0%  |     5343  |  173.00 |  3.6% |line.694
3| |   7.7%  |    42252  | 1991.25 |  5.1% |line.709
3| |  36.8%  |   202381  | 7241.50 |  3.9% |line.737
3| |   4.2%  |    23233  | 3554.75 | 15.2% |line.1129
3| |   9.0%  |    49446  | 15207.88 | 26.9% |line.1141
=============================================================
```

The issue that the application hits is load imbalance which is due to omp_set_lock and omp_unset_lock calls, which guard a critical region where only one thread at a time can enter. Given this requirement, the scaling that is achieved is very good. This load imbalance comes from line 1141 and line 684 of the major computational routine. In the code, we see the beginning of a lock–unlock region. Since this is a critical region, only one thread can enter the region at a time. This causes the other threads to wait at this lock. The variation in wait times for each of the threads causes the load imbalance (Figure 8.13).

```
for(i = 0; i< nng0; i++)
   {
      a2 = (*atomall)[natoms*o+i];
      /* add the new components */
#ifdef _OPENMP
      omp_set_lock(&(a2->lock));
#endif
```

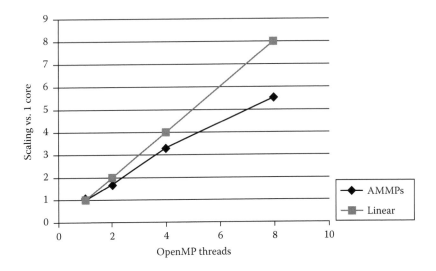

FIGURE 8.13   Scaling of AMMPs from 1 to 8 threads.

```
v0 = a2 -> px - a1px;
v1 = a2 -> py - a1py;
v2 = a2 -> pz - a1pz;
v3 = sqrt (v0*v0+v1*v1+v2*v2);

if (v3 > mxcut || inclose > NCLOSE)
  {
      r0 = one/v3;
      r = r0*r0;
  o      o      o
```

8.11 SUMMARY

In these eight benchmarks from SPEComp_2006, we have seen the factors that impact the performance when running OpenMP applications on a multicore socket. These results would vary to some degree on different multicore sockets; however, the underlying principles would still have the largest impact on the overall performance.

First and foremost, the application developer should strive to achieve the largest granularity possible, as in the example AMMP. Even when parts of the parallel construct are memory bound, if the granularity is large enough, the threads will naturally get out of synchronization in a large loop so that not all the threads access memory at the same time.

This natural load balancing is only possible when the granularity is large enough.

In all these applications, a very high percentage of the computation is parallelized with OpenMP constructs. This is not always that easy and should not be overlooked when examining other applications.

A prerequisite to good OpenMP, when one does not have large granularity, is good cache utilization. When the OpenMP region contains a few instructions <100/iteration, then the cache behavior dictates the performance achieved with OpenMP.

EXERCISES

1. What are the major characteristics of a good OpenMP application?
2. How does memory use affect performance on 8 threads?
3. Why might efficiency be better for 4-way OpenMP than 8-way OpenMP?
4. What is false sharing and how can one remove it?
5. What is load imbalance and how might one address loops that have load imbalance?
6. How does synchronization affect OpenMP performance?
7. When might a barrier be needed within a OpenMP region?
8. Explain the difference between STATIC and DYNAMIC scheduling? When would each be used?
9. In the APSI application, what might be a better way of handling the reduction variables WWIND and WSQ?

CHAPTER 9

Accelerators and Conclusion

Today, customers of very large HPC systems see that the electricity to run the system costs more than the purchase price of the hardware over the life of the massively parallel processor (MPP) systems. Fewer HPC sites can afford to supply the power required by some Petascale systems. Indeed, the biggest challenge to future Etascale systems is the amount of power required to run the system. Today all the chip manufacturers are supplying energy-efficient versions of their top performance chips. The Blue Gene® series of HPC systems are built upon a concept of using many less powerful processors that are more energy efficient. An evolving measurement of HPC systems is FLOPS/WATT; that is, the peak floating point operations supplied by an HPC system divided by the power required to run the system. The approach of using many less power-hungry processors does introduce more complexity in the scaling of applications across the machine. If there are more less-powerful processors, an application must use many more to achieve a performance as good as the previous generation system. Millions of processors in the next-generation HPC system are not inconceivable.

A second approach at energy-efficient processing is to use accelerators such as general purpose graphics processor units (GPGPUs) which dedicate all of their circuitry to parallel single instruction, multiple data (SIMD) processing elements and parallel memory channels. Given such an approach, the FLOPS/WATT measurement of many GPGPUs are very impressive. So an alternative to millions of low-performance processors would be thousands of nodes with accelerators. While the accelerator-based system does reduce the scaling problems imposed by having to run across millions of processors, they do introduce a programming challenge of their own. Today accelerators do not share memory with the cores on the node that comprise the MPP. While future accelerators are being designed that would have their own scalar processor, for the foreseeable future the user must contend with running their applications across two disparate systems which do not share memory. In addition to the issue of transferring data back and forth to the accelerator, the code that runs on the accelerator needs to be highly parallel with large granularity and must be able to vectorize at the low level in order to effectively attain a high sustained FLOPS/WATT.

9.1 ACCELERATORS

Consider the diagram of the latest Nvidia® GPGPU, the Fermi™ system [15] in Figure 9.1.

In this system, there are 16 Multiple Instruction, Multiple Data (MIMD) processing elements called streaming multi-processors (SM); within each SM, there are 16 streaming processors (SP) which are SIMD processors. There are also parallel read/write paths to and from dynamic random

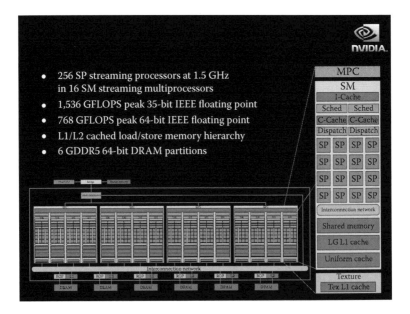

FIGURE 9.1   Fermi GF100 architecture diagram.

access memory (DRAM). For the accelerator processors to operate on data, it must be contained within the accelerator memory. The movement of data between the multicore socket on the same node as the accelerator must be performed over PCI links, and the bandwidth afforded by these links is typically not high enough to keep up with the processing power of the GPGPU. We will see that part of the porting process will be to minimize the amount of data that is moved between the two processing systems.

9.1.1 Using Extensions to OpenMP™ Directives for Accelerators

One way of looking at utilizing accelerators is to consider a hybrid MPI/OpenMP™ application where the OpenMP regions are run on the accelerator. The idea would be to run the MPI tasks on the multicore socket and set a barrier prior to the OpenMP regions, transfer necessary data over to the accelerator and then execute the kernel on the accelerator. After execution of the kernel control can come back to the multicore processor for continuing the MPI program. It will be important to identify a large OpenMP region with multiple inner DO loops to supply significant parallelism and inner loops should be vectorizable. Taking this concept one step forward, PRIVATE data in the OpenMP region would be the data that can be initialized and used on the accelerator without need to transfer from the multicore socket. SHARED data that are only used would be the data that would be transferred from the originating process on the socket to the accelerator prior to being used.

In particular, optimization would leave as much of this type of SHARED data on the accelerator as possible. SHARED data that are set within the OpenMP region would probably need to be transferred back to the multicore socket after the OpenMP region. Some shared data may have to be transferred to the accelerator prior to the execution of the kernel and back to the socket after the execution of the kernel.

There are numerous optimizations that can be applied to this concept. For example, the MPI tasks could execute in parallel with the execution of the accelerator OpenMP region. The MPI tasks could actually perform some communication with the other nodes and update the memory within the accelerator with the data from received messages. Of course, the application developer would have to assure that the multicore socket and the accelerator are synchronized at appropriate places.

Using an OpenMP structure as discussed has gathered significant momentum in the HPC community and the OpenMP committee is considering a set of extensions to allow the use of accelerators. In addition to the standard OpenMP directives, there would be directives for

1. Specifying what data need to be transferred to the accelerator prior to execution of the kernel.

2. Specifying what data need to be transferred back to the multicore socket after execution of the kernel.

3. Specifying what data will be persistent on the accelerator across OpenMP regions.

4. Specifying where data on the accelerator will be updated by data on the multicore socket.

5. Specifying where data on the multicore socket will be updated by the data on the accelerator.

This is just a subset of the directives being discussed; however, they give an idea of the type of control the programmer would have with such directives.

Given these extended directive sets, several compilers would be available to take the "Accelerator Region" and generate accelerator code and perform the necessary data motion. The intent would be to allow the programmer to develop an application for the use of an accelerator which would also be able to run on multicore systems without accelerators using OpenMP. While such a capability reduces the effort to utilize the accelerator, there is still a significant amount of work to do to assure that the code that runs on the accelerator will run faster than running only on the multicore socket.

9.1.2 Efficiency Concerns When Using an Accelerator

The largest concern is that of transferring the data back and forth between the multicore socket and the accelerator. Consider the OpenMP region as the specification of a data region. Within that region, data can be transferred back and forth between the accelerator and the multicore system. Within the data region there are accelerator kernels that consist of parallel, vector structures. Not all data need to be transferred to the accelerator prior to the execution of the first kernel. One of the optimizations would be to transfer enough data to execute the first computational kernel and then while the computation proceeds on the accelerator, the data required by subsequent kernels can be transferred asynchronously with the execution of kernels. To perform such an optimization, the data region must have a significant amount of work.

The second biggest concern is to have enough parallelism to utilize the MIMD processors (256 on the Nvidia Fermi system). If one outer DO loop does not have sufficient iteration count to cover that number of MIMD processors, one must consider collapsing two outer loops into one larger loop. Since OpenMP 3.0 has multilevels of parallel DO, this should be feasible to do with directives. Having much more than 256 parallel threads is a good thing, since the MIMD threads can be multithreaded to hide the memory latency within the accelerator. The actual target for this outer shared memory parallelism should be on the order of 1000–5000.

The next biggest issue is the vectorization of the inner loops and the memory accessing which those loops contain. Ideally, the innermost loops are vectorizable and the loops access data contiguously from memory. On the accelerator, memory latencies are relatively long and just as on a cache-based system, a contiguous chunk of memory is transferred to the local memory of the SP processors. When striding and indirect addressing is used in the inner loops, the effective memory bandwidth is poor and degrades the performance of the SP units.

9.1.3 Summary for Using Accelerators

This is meant to be a brief introduction to the use of accelerators and to show that the techniques contained in the chapters on Vectorization and OpenMP can be used to develop a good accelerator code. When addressing the issue of whether a particular application has potential for execution on an accelerator, one must examine the application at a high level to identify a data region which contains sufficient computational kernels to amortize the time to transfer the data back and forth to the accelerator. Consider the following steps for such an analysis:

1. Given a computational kernel, what data must be transferred back and forth between the accelerator and multicore socket? Compute the time required to transfer the data given the transfer rate of the PCI channel connecting the two systems.

   a. What data, if any, can be held in the accelerator memory?

   b. Can some of the data transfer be overlapped with computation?

2. Can multiple computational regions be combined into a single data region, so that less data need to be transferred?

3. Is there sequential, nonaccelerator codes that can be executed on the multicore host asynchronously with the execution of an accelerator kernel?

    a. Can MPI message passing be performed in parallel?

    b. Can I/O be performed in parallel?

9.1.4 Programming Approaches for Using Multicore Nodes with Accelerators

Today, the most common approach for utilizing attached accelerators is to write Compute Unified Device Architecture (CUDA™) and/or OpenCL kernels. CUDA is an extension of C and C++ to spawn and execute compute-intensive computation on the accelerator. Unfortunately, CUDA is not portable and is only usable on the NVIDIA GPGPU. Another similar approach that is more portable is OpenCL. The major complaint of both these languages is that they will not run on the multicore MPP without GPGPUs. Another approach pioneered by the Portland Group Compiler (PGI) is comment line directives that instruct the compiler to generate the kernel for the GPGPU, transfer the data to the GPGPU, spawn the kernel on the accelerator, and transfer the data back to the multicore host.

If this directive approach can approach the performance obtained by the CUDA code, the use of accelerators should expand significantly. With such a programming paradigm, an application developer can consider developing an applications that is structured to utilize MPI parallelism at a very high level (between the nodes on an MPP) with high granularity shared memory parallel structures that can either be run on the attached accelerator (if present) or run using the shared memory parallel cores on the node. Finally at the low level, the application is structured to utilize either the vector SSE instructions on the multicore host or the SIMD units on the accelerator.

This mechanism gives the application developer a performance portable application across current and future Petaflop and Exaflop systems.

9.2 CONCLUSION

We see a significant evolution in the HPC industry, comparable to that we witnessed in the 1990s when the "attack of the killer micros" moved us to distribute memory systems from simple shared memory vector processors. At that time, there was a tremendous programming challenge facing the application programmer to introduce some form of message passing to develop a program for a nonshared memory system. Many accepted the challenge and today applications are achieving performance gains that were inconceivable on shared memory vector systems. With the advent of new architectures to pursue the Exascale target, there are new challenges that would require significant commitment from the application programmer to port and optimize their application for the new systems.

With lower performance multicores, algorithms need to be scaled to multimillion, multicore processors, and significant work would be necessary to address the memory and network latency and bandwidth concerns.

With the advent of viable accelerators we essentially see a "back-to-the-future" situation where algorithms must be developed that can utilize large parallel vector systems with memory-accessing constraints.

One lesson we can derive from HPC history is that there is no "silver bullet." There is no software that automagically generates code. While software systems are available to help the application developer restructure their programs, the major work is performed by the programmer. From history, we can also revisit parallelization and vectorization techniques that were pioneered on machines such as the CDC™ Star, Cray 1™, Thinking Machines CM, Intel Paragon™, and Cray T3E™.

Unfortunately, there is a new generation of programmers in the HPC community who probably are not aware of the large repository of excellent algorithm research that was conducted on the early machines. This author often used research from the Illiac IV™ era to develop a parallel vector code for the Cray® vector systems. These new accelerators will significantly benefit from the work performed for the SIMD Thinking Machines CM5, Cyber 205™, NEC® SX, Fujitsu VPP, and Cray vector systems. They will also benefit from the work performed on the MIMD shared memory systems from Cray, NEC, and Fujitsu.

EXERCISES

1. Why is power consumption a concern with modern HPC?
2. What are the two approaches to lowering power consumption for HPC?
3. What additional information may be provided by directives for GPU execution?
4. What older HPC systems are similar to GPGPUs?
5. What are the major programming challenges for effectively utilizing GPGPUs today?
6. What will the major programming challenges for effectively utilizing accelerators that share memory with the host system?

Appendix A: Common Compiler Directives

T HE FOLLOWING TABLE GIVES compiler switch for the Portland Group compiler (PGI) and Cray Compilation Environment (CCE). Other compilers probably have similar options and the user is encouraged to examine appropriate manuals to obtain the options for their compiler. There are many more options than those that are given; this table primarily serves as an indication of the type of switches that are available.

Compiler Options for PGI	Compiler Options for CCE	Explanation
-tp target -mcmodel = small|medium	cpu = target_system	When cross compiling this option would indicate which processor you are targeting Small—objects less than 2 GBs Medium—objects greater than 2 GBs
Optimization Options -fast	*Default*	Both compilers have optimization for –O0 to –O3. The higher the level of optimization the more likely that answers may be less accurate. Default in both cases is two which is safe in most cases

continued

(Continued)

Compiler Options for PGI	Compiler Options for CCE	Explanation
Chooses generally optimal flags for the target platform. Use pgf95 -fast -help to see the equivalent switches. Note that this sets the optimization level to a minimum of 2	Optimization Level 2 CCE also have options to increase cache, scalar, vector and thread optimization separately by using. -cache$n$ -scalar$n$ -vector$n$ -thread$n$	
-Munroll$n$	-unroll$n$	Instructs the compiler to unroll DO loops
-Mipa = inline -Mipa = fast,inline	-ipa$n$	Instructs the compiler to inline routines. n = 1,2,3,4,5. Specifies the level of inlining
-mp	-omp	Instructs the compiler to recognize OpenMP™ directives
-Minfo	-rm	Instructs the compiler to generate information about what optimization was performed to the DO loop
The information comes out to standard error from the compiler	The information comes out in a .lst file	
-Mlist		Generate numbered source code listing
-g	-g	Debugging support

Appendix B: Sample MPI Environment Variables

THE FOLLOWING TABLE GIVES the environment variable for the Cray Message Passing Interface. While these are only valid on the Cray, it gives the user an idea of what message passing environment variable may be available on their system. Significant performance gains can be achieved by adjusting the environment variables available on the target system. Other environment variables may exist for other systems; so the user is advised to investigate what is available on their system.

Environment Variable	Explanation
MPICH_ENV_DISPLAY	This variable simply displays all of the MPI environment variable and their values. This is very useful to see what the defaults are if a variable is not set.
MPICH_MAX_SHORT_ MSG_SIZE	This variable sets a message size which dictates at which size in bytes; the receiver saves the message in a buffer versus instructing the sender to keep the message until the receive for the message has been posted by the application. Below this message the receiver will save the message in a buffer on the receive side.
MPICH_MAX_ VSHORT_MSG_SIZE	Specifies in bytes the size at which the message is included in the header.
MPICH_PTL_UNEX_ EVENTS	Specifies the maximum number of unexpected messages allowed. These are messages where the receive is not posted.
MPICH_UNEX_ BUFFER_SIZE	This environment variable specifies the size of the buffer to contain unexpected messages. Once again unexpected messages are those messages that are received prior to the application posting a receive.

continued

(Continued)

Environment Variable	Explanation
MPICH_RANK_ REORDER_METHOD	This environment variable allows for the user to specify the alignment of MPI tasks on the multicore node. Typically, there is SMP style where MPI tasks are allocated to a node sequentially. For example, an 8-process job launched on four dual-core nodes would be placed as NODE 0 1 2 3 RANK 0&1 2&3 4&5 6&7 For example, an 8-process job launched on two quad-core nodes would be placed as NODE 0 1 RANK 0&1&2&3 4&5&6&7 Folded rank For example, an 8-process job on four dual-core nodes would be placed as NODE 0 1 2 3 RANK 0&7 1&6 2&5 3&4 For example, an 8-process job on two quad-core nodes would be placed as NODE 0 1 RANK 0&3&4&7 1&2&5&6 Round Robin For example, an 8-process job launched on four dual-core nodes would be placed as NODE 0 1 2 3 RANK 0&4 1&5 2&6 3&7 For example, an 8-process job launched on two quad-core nodes would be placed as NODE 0 1 RANK 0&2&4&6 1&3&5&7 Additionally, there is an option to read a file that contains the alignment.
MPICH_ALLGATHER_ VSHORT_MSG	Used to specify when different algorithms are used for allgather.
MPICH_ALLGATHERV_ VSHORT_MSG	Used to specify when different algorithms are used for allgather.
MPICH_ALLREDUCE_ LARGE_MSG	Used to specify when different algorithms are used for addreduce.
MPICH_ALLTOALL_ SHORT_MSG	Used to specify when different algorithms are used for alltoall.
MPICH_BCAST_ONLY_ TREE	Used to specify when different algorithms are used for broadcast.
MPICH_MPIIO_HINTS	This includes a large number of tuning parameters for MPI I/O, such as striping factor, striping unit, and buffering parameters.

References

1. Wulf, W. A. and McKee, S. A., Hitting the memory wall: Implications of the obvious, *Computer Architecture News*, 23(1), 20–24, 1995.
2. Browne, S., Dongarra, J., Garner, N., Ho, G., and Mucci, P. A portable programming interface for performance evaluation on modern processors. *The International Journal of High Performance Computing Applications*, 14(3), 189–204, 2000.
3. L. Derose et al. Tools for high performance computing. *Cray Performance Analysis Tools*. Springer, Berlin, pp. 191–199, 2008.
4. Kanter, D. Inside Nehalem: Intel's future processor and system. http://realworldtech.com/includes/templates/articles.cfm; April 2, 2008.
5. Almasi, G. et al. Early experience with scientific applications on the blue gene/L supercomputer. *Euro-Par*, 560–570, 2005.
6. Camp, W. J., Tomkins, J. L. Thor's hammer: The first version of the Red Storm MPP architecture. In: *Proceedings of Conference on High Performance Networking and Computing*, Baltimore, MD, November 2002.
7. Baker, R. D., Schubert, G., and Jones, P. W., Convectively generated internal gravity waves in the lower atmosphere of venus. Part II: Mean wind shear and wave-mean flow interaction, *J. Atm. Sci.*, 57, 200–215, 2000.
8. Liu, S., Chen, J. H. The effect of product gas enrichment on the chemical response of premixed diluted methane/air flames. In: *Proceedings of the Third Joint Meeting of the U.S. Sections of the Combustion Institute*. Chicago, Illinois, USA, March 16–19, 2003.
9. Co-Array Fortran for parallel programming. Numrich, Reid, Rutherford Laboratory Report, RAL-TR-1998-060.
10. Carlson, W., Draper, J., Culler, D., Yelick, K., Brooks, E., and Warren, K., Introduction to UPC and Language Specification. CCS-TR-99-157, IDA Center for Computing Sciences, 1999.
11. DeRose, L. and Poxon, H. A paradigm change: From performance monitoring to performance analysis. *21st International Symposium on Computer Architecture and High Performance Computing*, Sao Paulo, Brazil, October 28–31, 2009, pp. 119–126.

12. Worley, P. Performance studies on the Cray XT. Cray Technical Workshop, Isle of Palms, February 24–25, 2009.
13. Kale, L. V. and Krishnan, S. CHARM++: A portable concurrent object oriented system based on C++. In: *Proceedings of the Conference on Object Oriented Programming Systems, Languages and Applications*, September–October 1993. ACM Sigplan Notes, Vol. 28, No. 10, pp. 91–108. (Also: Technical Report UIUCDCS-R-93-1796, March 1993, University of Illinois, Urbana, IL.) [Internal Report #93-2, March 93].
14. Mathis, M. M. and Kerbyson, D. J. A general performance model of structured and unstructured mesh Particle Transport Computations. *Journal of Supercomputing*, to appear, 2005. http://parasol.tamu.edu/publications/download.php?file_id=387
15. Streuer, T. and Stuben, H. Simulations of QCD in the era of sustained teraflop computing, Konrad-Zuse-Zentrum, ZIB Report 07-48, December 2007.
16. Rice, D. et al. SPEC OMP Benchmark Suite, December 2003, http://www.openmp.org.
17. *NVIDIA's Next Generation CUDA Compute Architecture—Fermi*. NVIDIA white paper V 1.1.
18. AMD's Magna-Cours' Server Chip Aims at Volume Market, Hachman, Mark, Article in PCMAG.com, 03-29-2010.
19. "Lustre: A future standard for parallel file systems?" invited presentation International Supercomputer Conference. Heidelberg, Germany, June 24, 2005.
20. Hennessy, J. L. and Patterson, D. A. *Computer Architecture—A Quantitative Approach*, Morgan Kaufmann publications, 2002.

Index

A

ABS, 109, 142
Accelerator, x, xiii, 30, 210–214
  data region 228–229
  directives, 212
Adaptive
  mesh, 78, 84, 176
  routing, 24
Add, 13, 101
Address, 4–5, 35, 42, 45–46, 160
  indirect, 113, 156
  translation, 5
Allocatable
  arrays, 27, 44, 203
AMD, xii–xiii, 5, 14, 16, 51
Amdahl's law, 63, 180
AMMP, x, 204–206
APPLU, ix, 192, 195
Aprun, 192
APSI, ix, 196–199, 213
ARMCI, 85
Array
  assignment, 32, 60–62, 66–67
  automatic, 27, 29, 65
  constant, 64
  padding, 28–29
  sections, vi, 39, 42, 49
Art, ix, 203–204
Associativity, xi, 7–9, 16

B

Barrier, 23, 81, 83, 89, 108, 165, 193–194,
    207, 211
BLAS, 42, 134–136, 160, 189

Blue Gene®, 20, 23, 209, 221
Buffer
  application, 24, 59, 183–184
  look-aside, 4, 99
  messages, 59, 183–184, 219
  receive, 59, 78, 183–184, 219

C

C
  language, xii, 27, 39, 66–68, 83,
      103–106, 116, 139–143
  pointers, 27, 32
C++, xii, 39, 158, 214
Cache
  associativity, xi, 7–9, 16
  blocking, 12, 35–36, 39, 57, 60, 78,
      109, 145, 147–148, 155,
      159–160, 171
  level 1, 2, 5–10, 16–17, 40–41, 70,
      107–108, 147, 154, 190–191, 197
  level 2, 2, 7, 9–10, 16–17, 40–41, 49,
      147–148, 154–155, 189, 218
  level 3, 10, 15–17, 40, 148–149, 189
  line, 5–12, 16, 35, 68–70, 86, 107–108,
      110, 156, 191, 196–197, 200
  miss, 3
  reuse, 35–36, 47, 95, 108, 129, 145–146,
      156, 191
CAF. *See also* Co-Array Fortran, vii, xiv,
    77–80, 83–86, 86, 170, 221
CDC™. *See also* Control Data
    Corporation
  star, 40, 215
CHARM++, 176

Children, 79–82

Clock cycle, xii, 1–2, 11, 13–15, 30, 109

CM. *See also* Connection machine, 40, 164

Co-Array Fortran, vii, xiv, 77–80, 83–86, 170, 221

Collectives
communication, vii, 52, 56–57, 70, 78–79, 89
MPI, 56–57, 86, 179

Combining
algorithm, 52, 185
collectives, 57, 179
loops, 148, 152
messages, vii, 28, 58
parallel regions, 63

Comment line directives, xiv, 12, 32, 38–39, 63, 214

Common
block, 28, 43, 66, 106, 133–134, 157
subexpression, vi, 47–48

Commutativity, 47

Compiler
switches, vi, 32, 41–42, 217–218

Computational intensity, viii, 44, 94–95, 100–103, 107–108, 140, 142, 157, 159–160, 170, 202, 204

Connection machine, 40, 164

Contiguous addressing, 156

Control Data Corporation, 40

Controlled store, 33

Counters hardware, v, 3, 6, 43, 94, 106–107, 202–203

Cray
CCE, xiv, 106, 112, 114, 119, 125–146, 217–218
Inc, xii, 55, 89–90
research, 32
T3E, 215

CrayPat™, 3, 44, 55, 89, 94, 107, 170, 177, 202, 204

Crossbar, 20

CUDA™, 214, 222

D

Decomposition, vii, 11, 26, 49, 52–54, 57, 61, 63, 72–73, 75–76, 86, 88, 161–162, 174–176, 179, 181

Dedicated mode, 19–20, 23, 104, 108

Dependency, vi, 13, 31, 49, 64, 111, 156, 194

Derived
metric, 3–4, 94, 107
type, vi, 43–44, 49, 203

DGEMM, 140, 142–144, 196

Distributed memory, xi, xiv, 31, 51, 66

DOT_PRODUCT, 42, 116–117, 139

Double precision, 13, 140

DRAM, memory, 2–3, 210–211

E

Environment variables, vii, x, xiv, 61, 219–220

EQUIVALENCE, 28

Exascale, 215

Exponents, vi, 46–47

F

False sharing, vii, 16–17, 68–69, 196, 207

Fast Fourier Transform, 57, 179

Fermi™, 210, 213, 222

FFT. *See also* Fast Fourier Transform, 57, 70, 78–79, 179–180, 185

Finite difference, 20, 52, 88, 156, 162, 189

FLOPS, xi, xvi, 1, 10, 14, 44, 94, 142

FLOPS/WATT, 209–210

FMA3D, 202

Fork, 71

Fortran
2003, vi, 49, 221
module, vi, 28–29, 39, 43, 64–66, 135, 157
scoping, 64

Fragmentation, 5, 28

Functional
decomposition, 72–73, 76–77
units, xi-xiii, 1–2, 11–13

G

GALGEL, ix, 194–196

Garbage collection, 24, 28

GASNET, 85

Gather, 34, 55, 62, 85–90, 110

Get, 77–79, 83–86

Global Sum, 23, 26, 79–83, 164–165, 185
GPGPU, xv–xvi, 25, 30, 49, 112–113, 160, 209–211, 214–215
Gprof, 88
Granularity, 63, 66–68, 73–74, 187, 190, 194, 203, 206–207, 210

H

Halo exchange, 26, 58, 61, 165
Haperstown, 17
Hardware counters, v, 3, 6, 43, 76, 94, 106–107, 202–203
Hennessy, 4, 222
Hybrid Programming Paradigm, vii, ix, xiv–xiv, 52, 70, 169, 180–183, 187, 211
HyperTransport™, 15–16

I

I/O, ix, xv, 19–29, 88–89, 96, 182–185, 211, 214, 220
IBM®, 20, 23
IF statements, vi, viii, 30–34, 69, 111, 118–128, 132, 135–136, 158, 198
Illiac IV™, 215
Index reordering, viii, 157
Indirect addressing, vi, 14, 30, 34, 84–85, 101, 110, 112, 130, 155–156, 213
Injection bandwidth, xiv, 22, 24, 26, 62, 176
Instrumentation, 90, 165
Intel Paragon™, 215
Interconnect bandwidth, 19, 57, 176, 180
Interprocedural analysis, vi, 35, 38–39
Intra-node, 85
IPA. See also Interprocedural analysis, 38, 218
IVDEP, 32, 111, 113

J

Java, xii
Job placement, v, 21

K

Kanter, 15, 221

L

LAPACK, 42
Latency
    memory, xiii–xiv, 1–2, 12, 213
    network, 20, 22, 26, 36, 52, 55–58, 85–86, 180
Leslie3D, ix, 88, 148, 162–163
Level
    1 cache, 2, 5–10, 16, 40–41, 70, 107–108, 147, 154, 156, 171, 190–191, 197
    2 cache, 2, 7, 9–10, 16–17, 40–41, 49, 147–148, 154–155, 189, 218
    3 cache, 10, 15–17, 40, 148, 154–155, 189
Light weight operating system, 23
Link
    bandwidth, 22, 24, 42, 162
    errors, 22
Linux®, 4, 23–24, 51, 183
Load
    balancing, 55, 73–74, 176, 207
    imbalance, ix, xii–xiii, 3, 56, 59, 61–62, 66, 88–92, 97, 162, 165–169, 176–181, 185, 200–207
Local data, 29, 39
Lock, 40, 76, 108, 205
Logical
    address, 4–5, 12
    memory, 4–5, 12, 29
    page, 4
Loop
    carried dependency, 31, 113
    fission, iix, 157
    jamming, iix, 159
    reordering, iix, 151, 156–157
    splitting, iix, 124, 157
    tiling, 35, 145
    unrolling, vi, iix, 35–39, 138, 143, 157–160
Lustre, 183, 222

M

Malloc, 27, 160
Master thread, 63, 196
MATMUL, 42
MAX, 125, 158

McKee, 3, 221
Memory
  alignment, v–vi, 10–11, 14, 28, 34, 36,
    49, 183, 220
  allocation, vi, 11, 21, 24, 27–28, 59,
    66, 176
  architecture, v, xiii, 1–2
  bandwidth, vii, xii, xv, 2, 16, 24, 26,
    70–71, 100–102, 104, 108–109
    124, 153, 156, 176, 181,
    189–192, 213
  banks, 6, 10–11
  contention, 73
  controller, 12, 15
  fragmentation, 6, 28
  hierarchies, 1
  logical, 4–5, 29
  management, v, 24
  physical, 4–5, 160
  prefetching, v–vi, xiii, 11–12, 36–37,
    49, 79, 109
  wall, v, 2, 221
Message
  passing, vii, xii–xv, 19–20, 23, 51, 55,
    59, 78, 85, 88–89, 161, 167,
    179–180, 214, 219
  rate, 24
  transfer time, v, 22–24, 53, 58–59, 79,
    81, 84, 183, 211–214
MGRID, ix, 145, 147, 151, 192–193
MIMD, xi, xiii, 210, 213, 225
MIN, 125, 158
MODULE Fortran, 28, 43, 64–65,
  106, 157
Moore's law, xi
MPI, task placement, 21
MPI_ALLREDUCE, 56–57, 83, 161,
  165–166, 177, 179
MPI_ALLTOALL, 22, 52, 56, 78–79, 84,
  161, 179–180
MPI_ISEND, 55–56, 92–96, 165,
  166–168
MPI_REDUCE, 56, 57, 163, 179
MPI_SENDRECV, 59
MPI_VGATHER, 56, 179
MPI_VSCATTER, 56, 179
MPI_WAIT, 59, 90–96, 163–168
MPI_WAITALL, 90–91, 96, 165–168

MPP, v, viii, xiii–xiv, 19–25, 51–52, 61,
  76–77, 79, 81, 83, 95–99, 102,
  173, 183, 209–210, 214, 221
Multiply, 12, 55, 116, 140, 158
Mutex, 76

N

NAMD, 176
NASA parallel benchmark, 145, 192
Nearest-neighbor, 10–12, 24, 60–61, 78,
  162, 166, 171–172, 176, 185
NEC®, 215
Nehalem, 14–17, 221
Nested DO loops, vi, 35–37, 140, 143, 145,
  148, 153
Network interface computer (NIC), v, xiv,
  24, 77
Nine point stencil, 164–165, 169
Non-reproducibility, 29
NPB. *See also* NASA parallel benchmark
Nvidia®, 210, 213–214, 222

O

Oak Ridge National Laboratory, 167
OpenCL, 214
Operating system
  demons, 23, 45, 63
  jitter, 19, 23, 26
Opteron™, 14–15, 100, 147–148, 191
Overhead
  CAF
  CrayPat, 44, 90
  OpenMP™, vii, 63, 66–68, 71–74, 84
  vectorization, 30–31, 34, 109, 116

P

Packed mode, 104
Padding, 28–29
Page
  logical, 4–5, 12
  physical, 4–5, 12, 27–28, 99
  size, 4
PAPI, 3
Parallel
  files, ix, 183–184, 222

sections, 62
  threads, 61, 77, 213
Particle in cell, 78, 88, 156
Pattern matching, 143–144
Patterson, 4, 222
PCI bus, 16, 211, 213
Pencil decomposition, 53
Physical
  memory, 4–5, 12, 160
  page, 5
Pipelining, 72–73, 77
Plane decomposition, 26, 100, 111, 146, 175
Pointer, 28, 90
Point-to-point communication, vii, ix,
      52, 57–60, 180, 185
POP, ix, 40, 151, 164, 166, 169
Portland Group, xiv, 36–37, 214, 217
Posix® threads, vii, 71
Pre-posting receives, vii, 24, 58–59, 86, 167
Private data, 43, 55, 64–69
Put, 77–85

R

Reciprocal approximation, 151
Registers, 2, 13, 30, 48, 110, 143
Remote
  get, 77–79, 83–86
  put, 77–85
Runge Kutta, 161–162
Runtime statistics, vii, xiv–xv,
      49, 61, 87

S

Sampling, 90–91, 205
Scalar promotion, viii, 157
Scaling
  strong, 56, 57, 60, 77–78, 161–162, 164,
      181, 185
  study, 56, 88, 95, 171–172, 175
  weak, 54, 56, 60, 87–89, 95, 97, 162,
      171–172, 175
Scatter, 33–34, 85–86, 110, 202
Scheduler, 20–22, 71
S3D, ix, 42, 54, 67, 89, 195,
      171–172, 175
Sequence association, 29

Shared memory aware MPI, 52
SHMEM, 85
SIGN, 158
SIMD, xi, xvi, 40, 209–210,
      214–215
SPEC®
  SPEC_OMP, 43, 69
SPMD, 55
SSE
  SSE2, 13
  SSE3, 13, 109–114, 119,
      129–130, 160
Star 100, 40, 215
Statement function, 130, 144–145
Static arrays, 27, 66, 207
Storage association, 29
Strength reduction, vi, 44–45, 49
Strip mining, viii, 128, 153–154, 158
Strong scaling, 56, 57, 60, 77–78, 161–162,
      164, 181, 185
Subroutine
  arguments, 62, 106
  inlining, viii, 158, 218
Superlinear scaling, 41, 162, 164,
      177, 185
Surface area, 53, 174–175

T

TASKCOMMON, 43
Thinking Machines, 40, 215
Thread private, 43
3D, Transpose 24, 56–57, 70, 171,
      179–180, 185
Tiling, 36, 145
TLB
  data misses, 5, 100, 107, 170
  thrashing, 5, 16
Topology, v, 19–21, 26
Torus, 20–22, 26
Totalview, 77
Translation look-aside buffer. *See also* TLB

U

Unexpected message buffer, 219
University of Tennessee, 3
Unpacked mode, 104–109

V

Variable scoping, vii, 64, 86
Vector
    dependency, vi, 13, 31, 49, 64, 111,
        156, 194
    routine, 135
Vectorization
    overhead, 30–31, 34, 109, 116
Volatile, 80

W

Wait time, 89, 109

Weak scaling

Weak scaling, 54, 56, 60, 87–89, 95, 97,
        162, 171–172, 175
Worley, Pat, 167, 222
Wulf, 3, 221
WUPWISE, 9, 135–136, 187–188, 190

X

x86, xii, 110, 112, 114, 119, 130, 139
XT, 20, 23, 61, 183, 192, 222

Z

Zeon, 14